Sean KRAMER

FIELD GUIDE
For Quick Identification

OUTDOOR PLANTS™ OF THE SOUTHWEST

Archibald W. Roach, Ph.D.
**Professor Emeritus,
North Texas State University**

W9-DHX-057

Published by arrangement with
the Jones-Kenilworth Company, Inc.

**TAYLOR PUBLISHING COMPANY
Dallas, Texas**

art director
Gloria Cochran

cover design
Linda Daboub

leaf illustrations
A. W. Roach, Ph.D.
Linda Daboub

Library of Congress Catalog Card Number: 81-85659
ISBN: 0-87833-323-1

Printed in the United State of America

Preface

The thrill and excitement of being able to point to and name hundreds of Nature's plants is second only to having the knowledge and the ability to use this skill. Plants are our true natural environment. You can begin your discovery in your own yard.

My years of Scouting and love of Nature, shared with countless other people, prompted me to suggest the idea for this book to my long-time friend, Archibald W. Roach, Botanist and Professor Emeritus of North Texas State University.

In addition to the wealth of knowledge offered in this handbook, Dr. Roach introduces the "simplified scientific method of determination," which ensures the challenge of discovery and fun for all. He uses non-technical language and pictures to help you find the common name of a plant unknown to you. This is the Scouting method which lets your fingers and eyes go with your walking, and learn and enjoy as never before.

Traditional taxonomy has relied on fruit and flowers for identification and classification of plants; however, plants are in their fruit and flower conditions only occasionally and only for a short period of time. Also, most keys to identification are replete with scientific terminology, understood only by a trained botanist. Dr. Roach has prepared this key in a unique format, which allows all readers who view with care, to properly identify plants.

Edward T. Farris, B.A., M.A., D.D.S.
Scout Master – Silver Beaver

WARNING

Before handling woody specimens, become familiar with the two plants shown here. They are poisonous to the touch for most people, producing painful blistering and swollen tissues.

They are from left to right: POISON SUMAC and POISON OAK/IVY. Read page 35 which shows and tells you how to determine if leaves are compound. The leaves of these dangerous plants are compound. *Poison Sumac* grows in wet places, its compound leaf bearing 7-13 leaflets laterally and having red leaflet stalks and veins. *Poison Oak/Ivy* bears its leaflets in clusters of threes; its veins are greenish. The leaf edge may be smooth or notched as shown. *Poison Ivy* is commonly a low forest shrub but may also grow as a vine. **DO NOT HANDLE THESE PLANTS.**

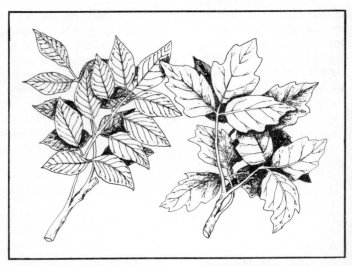

Poison Sumac   **Poison Oak/Ivy**

How to Use This Picture Guide

All of the native woody plants, trees, shrubs, and vines, of Texas, Arkansas, Oklahoma, Louisiana and New Mexico are included in this identification manual. You follow a key as you would in a treasure hunt. At points along the route, it asks you to look at your tree, shrub or vine, read the illustrated clues and decide which picture to follow next. You find the treasure whenever the route you are following ends in a name — the name of the plant with which you are unfamiliar.

All pictures and descriptions are based on vegetative characteristics because plants are not always in bloom and thus floral pictures would be roadblocks to their easy identification.

The scientific name, in italics, follows the common name. It is the only guaranteed world-wide name for a species. A species may have many common names in different languages, or several different species. To avoid confusion, we have selected a common name for each species that we think is the most appropriate for our five state area.

All species are listed in alphabetical order in the index.

An Example of How to Use the Key: Turn to the first page of the key. First you determine: (1) does the specimen possess leaves, or (2) doesn't it? The clues are arranged to the right of the picture. There may be two, three or four clues on a page. Always read all the clues before deciding which one will lead you to the next fork in the path. If your specimen lacks leaves, look for the picture to the right of the leafless clue, also marked Key Illustration, for example (KI) 2, which leads you to the left-hand margin on page 3. On page 3, your new clues are: does it have thorns or doesn't it? If it doesn't, look for that picture (KI) 5 in the left-hand margin of page 9. You are now at the end of the trail and the two choices end in *names* with their pictures marked as *figures* and not as key illustrations. Your specimen is either EPHEDERA or WAX EUPHORBIA. If it lacks branches, it is WAX EUPHORBIA. Now, read the description to the right of the picture, Figure 7, and sure enough, it is on a limestone hill, in the Big Bend of Texas — the only place it grows.

PICTURE GUIDE

1. If your specimen lacks leaves, that is, if the stem "joints" are bare, as shown to the right, go to this picture (KI 2), in the left-hand margin of page 3.

```
┌─────────────────────────────┐
│                             │
│                             │
│                             │
│        START HERE           │
│                             │
│                             │
│                             │
└─────────────────────────────┘
```
KI 1

2. If your specimen possesses leaves at the stem "joints" from mere scales up to broad, flat shapes, as shown to the right, find this picture (KI 6), in the left-hand margin of page 11.

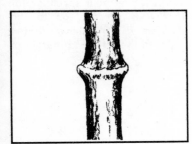

KI 2

KI 6

1. Look over your specimen for thorns. If present, go to the diagram (KI 3), shown to the right, in the left-hand margin of page 5.

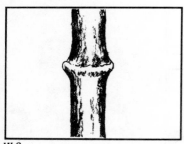

KI 2

2. If your specimen lacks thorns, find the picture shown at the right (KI 5), in the left-hand margin of page 9.

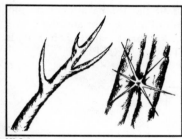

KI 3

KI 5

1. If the thorns look like those in the picture to the right and the young stems are smooth, it is CRUCIFIXION THORN.

2. If the thorns appear as pictured at the right and the young twigs have short hairs, it is TEXAS ADOLPHIA.

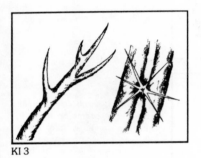

KI 3

3. If the thorns are short, frequently curved upward and borne on whip-like stems, it is OCOTILLO.

4. If the thorns are spines and the plant usually fleshy, your specimen is a woody cactus; go to the symbol (KI 4), shown to the right, in the left-hand margin of page 7.

5

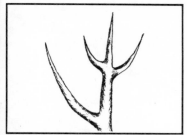

Fig. 1

CRUCIFIXION THORN
(Koeberlinia spinosa)
This leafless shrub or small tree looks like a tangled mass of green thorns and grows on the dry plains and hills of southwestern Texas and southern New Mexico. It is a member of the caper family *(Capparidaceae).*

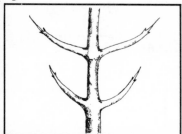

Fig. 2

TEXAS ADOLPHIA
(Adolphia infesta)
A shrub up to 6 ft. tall, leafless, densely covered with green thorns. It is found in the dry mountains of western Texas. Adolphia belongs to the buckthorn family *(Rhamnaceae).*

Fig. 3

OCOTILLO
(Fouquieria splendens)
A spidery shrub of slender, groved spiny stems up to 20 ft. high. The small, leathery leaves fall away early in the spring after the flaming red flowers are produced. It grows on rocky flats or hillsides in the deserts of southwestern Texas and New Mexico. It belongs to the candlewood family *(Fouquieriaceae).*

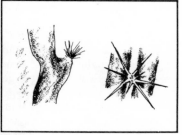

KI 4

1. If the thorns are short, it is DEERHORN CACTUS.

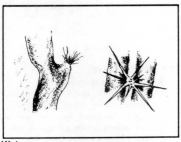

KI 4

2. If the thorns are long, it is a CHOLLA.

Fig. 4

DEERHORN CACTUS
(Cereus greggii)
A sprawling, slender, few-branched, fleshy shrub that is rarely over 3 ft. tall. It is found in dry mountains at altitudes of 2,000-4,000 ft. in western Texas and New Mexico. It belongs to the cactus family *(Cactaceae)*.

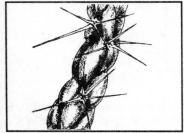

Fig. 5

CHOLLA
(Opuntia spp.)
Some seven species with cylindrical rather than "earlike" jointed stems are called cholla in this genus. They are slender, branched and grow from 2 to 10 ft. tall in desert flats, hillsides and the dry plains of western Texas and New Mexico. The chollas belong to the cactus family *(Cactaceae)*.

1. If your specimen looks like a bunch of branched green sticks, it is EPHEDERA.

KI 5

2. If your specimen looks like a cluster of green rods, it is WAX EUPHORBIA.

9

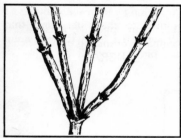
Fig. 6

EPHEDRA
(Ephedra spp.)
A similar group of eight species that consists of low green-stemmed shrubs with opposite branches and no leaves. They occur on the dry plains, mesas and mountains up to 6,000 ft. in western Texas and New Mexico. They belong to the ephedra family *(Ephedraceae)*.

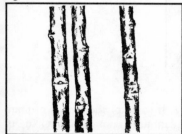
Fig. 7

WAX EUPHORBIA
(Euphorbia antisyphilitica)
A shrub under 3 ft. in height with rod-like, green, leafless stems. Found on limestone hills in the Big Bend area of Texas. It is a member of the spurge family *(Euphorbiaceae)*.

10

1. If the leaves of your specimen are long (non-scale-like) as shown to the right, find that drawing (KI 9) in the left-hand margin of page 17.

KI 6

2. If some of the leaves of your plant are minute or scale-like, as shown to the right (KI 7), go to the left-hand margin of page 13.

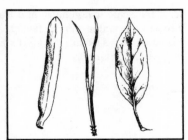

KI 9

KI 7

1. If the scale is hairy, as shown to the right, it is KRAMERIA.

2. If the scale is non-hairy and a broadish triangle on a fleshy, jointed stem, it is PICKLEWEED.

KI 7

3. If the scale is non-hairy and minute, go to the drawing (KI 8), shown to the right, in the left-hand margin of page 15.

13

Fig. 8

KRAMERIA
(Krameria spp.)
This complex of three similar species are low, thorny shrubs up to 2 ft. high in our area. The pea flowers are quite irregular and the legumes prickly. They inhabit dry rocky slopes and plains of western Texas and New Mexico. They belong to the pea family *(Leguminosae).*

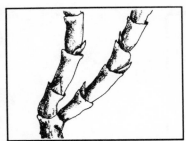

Fig. 9

PICKLEWEED
(Allenrolfea occidentalis)
This erect shrub with succulent branches grows up to 5 ft. tall. It inhabits alkaline flats and valleys in western Texas and New Mexico. It belongs to the goosefoot family *(Chenopodiaceae).*

KI 8

14

1. If the scales have glands on their backs and are aromatic when crushed in your hand, it is a JUNIPER.

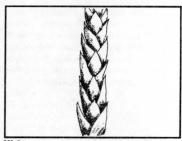

KI 8

2. If the scales are non-glandular and non-aromatic, it is TAMARISK.

JUNIPER

(Juniperus spp.)

Fig. 10

The junipers consist of many species growing as low shrubs up to trees 50 ft. tall. They are commonly called cedars because of their aromatic heartwood. They all produce pistillate cones that look like small berries. The VIRGINIA JUNIPER grows in the eastern states of our area; whereas, the most prevalent juniper to the west is the WESTERN JUNIPER. The juniper of central Texas is ASHE'S JUNIPER and the MOUNTAIN JUNIPERS are the common juniper, varieties of which may form extensive patches under 2 ft. tall. Junipers are gymnosperms and belong to the pine family *(Pinaceae)*. The ARIZONA CYPRESS *(Cupressus arizonica)* has the same type of minute, glandular and scale-like foliage as do the junipers. Its 1 in. long cones, however, are woody, dry and consist of 6-8 shield-shaped scales. This tree ranges from 40-90 ft. tall and grows on dry slopes in the Chios Mountains in Texas and southern New Mexico. It belongs to the cypress family *(Cupressaceae)*.

TAMARISK

(Tamarix gallica)

Commonly called SALT-CEDAR in our area, this scale-leaved twisted shrub or small tree forms extensive thickets along rivers and creeks in the western states or in saline soils along our southern coasts. When it blooms, it forms plume-like clusters of small, pink flowers at the ends of its branches. It belongs to the tamarisk family *(Tamaricaceae)*.

Fig. 11

16

1. Decide if the leaves of your specimen are much longer than broad (less than ⅛ in. wide) and resinous in odor when crushed in your hand. If so, look for the symbol (KI 10), in the left-hand margin of page 19.

KI 9

2. If the leaves are relatively broad (more than ⅛ in. wide) or if less than ⅛ in. wide and non-resinous-fragrant, proceed to the drawing, shown to the right (KI 18), in the left-hand margin of pages 35 and 35A.

KI 10

KI 18

1. If the leaves are needle-shaped and clustered as shown to the right, look for this symbol (KI 11), in the left-hand margin of page 21.

2. If the leaves are flat, round-tipped and non-clustered, as shown to the right, go to this symbol (KI 17), in the left-hand margin of page 33.

KI 10

3. If the leaves are needle-shaped, sharp-tipped and non-clustered, look for the symbol shown to the right (KI 16), on page 31.

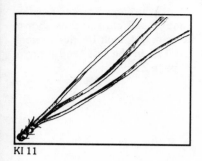

KI 11

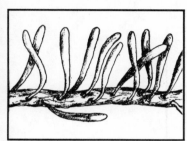

KI 17

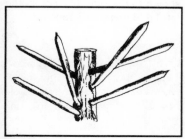

KI 16

1. Count the needles, if there are 2 or 3 per cluster, go to (KI 13), in the left-hand margin of page 25.

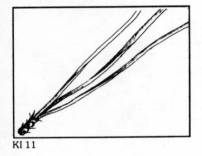

KI 11

2. If the needles are 5 per cluster, go to the drawing, shown to the right (KI 12), in the left-hand margin of page 23.

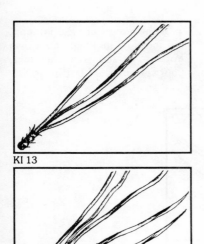

KI 13

KI 12

1. If the leaves are 2-3 in. long, but mostly 3 in. long, and a twig appears as shown to the right, your specimen is LIMBER PINE.

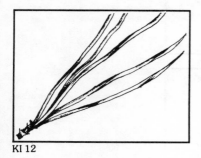

KI 12

2. If the leaves are 1-2 in. long, but mostly 2 in. long, and a twig appears as shown to the right, your specimen is BRISTLE-CONE PINE.

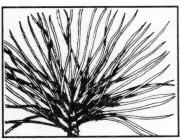

Fig. 12

LIMBER PINE
(Pinus strobiformis)
Gnarled trees growing mostly near timberline in the mountains of western Texas and New Mexico. The young branches have silvery-grey bark. The pistillate cones are 2-10 in. long. It belongs to the pine family *(Pinaceae)*.

Fig. 13

BRISTLECONE PINE
(Pinus aristata)
This pine occurs mostly as a low, wind-trimmed shrub near timberline in the mountains of western Texas and New Mexico. The cones are 2-3 in. long with their scales terminated by re-curved prickles. It belongs to the pine family *(Pinaceae)*.

1. If the leaves of your specimen are always clustered in bundles of 3, it is LONGLEAF PINE.

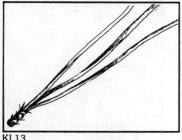

KI 13

2. If the leaves of your specimen are in bundles of 3's but sometimes 2's, go to (KI 14), in the left-hand margin of page 27.

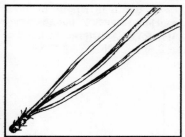
Fig. 14

KI 14

LONGLEAF PINE
(Pinus palustris)
A few-branched pine growing up to 125 ft. tall on deep sand in southeastern Texas and southern Louisiana. Its reddish-brown cones are 6-12 in. long. It belongs to the pine family *(Pinaceae).*

1. If the leaf length of your specimen is ¾ to 1¾ inches, it is a PINYON PINE.

2. If the leaf length of your specimen is 3-5 inches, it is SHORTLEAF PINE.

KI 14

3. If the leaf length of your specimen is 5-10 inches, find the clue (KI 15), shown to the right, in the left-hand margin of page 29.

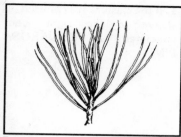

Fig. 15

PINYON PINE

(Pinus cembroides or *edulis)*
Short-trunked trees up to 50 ft. tall with low-spreading branches and roundish, small female cones. They occur at mid-altitudes in the mountains of western Texas and New Mexico. The two species intergrade and are weakly separable. They belong to the pine family *(Pinaceae).*

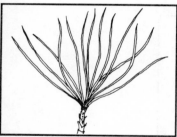

Fig. 16

SHORTLEAF PINE

(Pinus echinata)
The common upland pine of eastern Texas, eastern Oklahoma, Arkansas and Louisiana. It forms dense stands of trees up to 100 ft. tall with oval-shaped pistillate cones 1½-2½ in. long. It belongs to the pine family *(Pinaceae).*

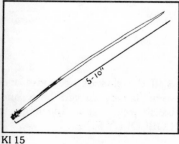

KI 15

1. If the clusters are crowded, as shown to the right, your specimen is PONDEROSA PINE.

2. If the clusters are loose and the leaves straight and shiny-green, your specimen is SLASH PINE.

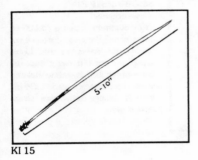

KI 15

3. If the clusters are loose and some leaves curved back and bluish-green, your specimen is LOBLOLLY PINE.

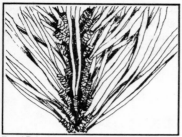
Fig. 17

PONDEROSA PINE
(Pinus ponderosa)
The large pine of our western mountains, attaining heights of over 200 ft. In our area, it occurs in the mountains of western Texas, New Mexico and northwestern Oklahoma. Its cones are oval and 3-6 in. long. It belongs to the pine family *(Pinaceae)*.

Fig. 18

SLASH PINE
(Pinus elliottii)
A fast-growing tree up to 100 ft. tall. It inhabits sandy swamps and stream banks in southeastern Louisiana in our area. Usually tapped for its turpentine-yielding resin. Its cones are quite variable but its leaves are more shiny-green than those of the other pines. It belongs to the pine family *(Pinaceae)*.

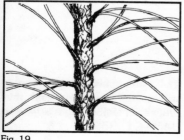
Fig. 19

LOBLOLLY PINE
(Pinus taeda)
Tall conifers (up to 150 ft.) of the lowlands of eastern Texas, Louisiana, southern Arkansas and southeastern Oklahoma. This fastest-growing of all pines is widely used for reforestation projects. It tends not to drop its elongate cones that are 3-5 in. long. It belongs to the pine family *(Pinaceae)*.

30

1. If the young branches (twigs) of your specimen are non-hairy, it is COLORADO BLUE SPRUCE.

2. If the young branches (twigs) of your specimen are hairy and the leaf has a groove on its upper surface, it is a DOUGLAS FIR.

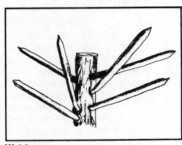

KI 16

3. If the young branches of your specimen are hairy and if the leaf has no groove, it is ENGLEMANN SPRUCE.

4. If your specimen grows in a swamp or a wetland and the branchlets look like feathers, it is BALD CYPRESS.

Fig. 20

COLORADO BLUE SPRUCE
(Picea pungens)
Evergreen, found in the mountains of New Mexico. The drooping female cones are light brown and about 3 in. long. It belongs to the pine family *(Pinaceae)*.

Fig. 21

DOUGLAS FIR
(Pseudotsuga menziesii)
An evergreen up to 200 ft. in height with 2-4 in. long female cones that bear distinguishing trifid bracts below the cone scales. It occurs in the mountains of Trans-Pecos Texas and New Mexico. It belongs to the pine family, *(Pinaceae)*.

Fig. 22

ENGLEMANN SPRUCE
(Picea engelmannii)
Straight-trunked evergreen up to 120 ft. in height of pyramidal form. Found in our area in the upper reaches of the New Mexico mountains. The drooping female cones are 1-2 in. long with tan scales. It belongs to the pine family *(Pinaceae)*.

Fig. 23

BALD CYPRESS
(Taxodium distichum)
A deciduous conifer up to 130 ft. in height with a swollen base made up of ridges. Female cone spherical, about 1 in. in diameter and yellowish-brown. It is a common inhabitant of swamps, occurring in our area in East Texas, Louisiana and as far north as eastern Oklahoma and Arkansas. It is a member of the pine family *(Pinaceae)*.

1. If the leaves are few and twisted, it is WHITE FIR.

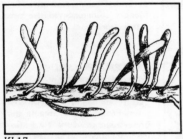

KI 17

2. If the leaves are many and never twisted, your specimen is ALPINE FIR.

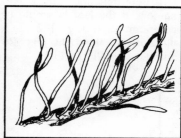

Fig. 24

WHITE FIR
(Abies concolor)
Montane evergreen conifer of the northern forests of New Mexico in our area. Young trees are conical, older are more rounded. Female cones are upright with crowded scales and are 3-5 in. long. It belongs to the pine family *(Pinaceae)*.

Fig. 25

ALPINE FIR
(Abies lasiocarpa)
An upper montane conical conifer, occurring in our area in the northern New Mexico mountains. Its upright cones are 2-4 in. in length with reddish-brown scales. It is in the pine family *(Pinaceae)*.

At this crossroads we have a major determination to make, namely, if the leaf is simple or compound. Study the two diagrams to the right (Fig. 26 and 27).

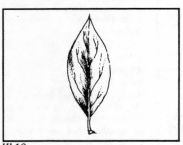

KI 18

The crucial point is the position of the bud, as indicated by X→. If there is only one blade past the bud in the angle of the leaf stalk, the leaf is simple; if there is more than one (then each is called a bladelet or leaflet), it is compound. Notice that there is no bud in the stalk angle of the bladelet as shown by O→. Also, bladelets may be arranged laterally, as shown above or terminally as shown in Fig. 28 to the right.

35

Fig. 26   Simple

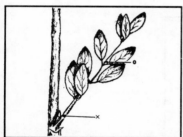

Fig. 27   Compound

Fig. 28

1. If your specimen possesses simple leaves and is a tree or shrub, start with the drawing shown to the right (KI 53), in the left-hand margin of page 105.

2. If your specimen possesses compound leaves, and is a tree or shrub, look for the drawing to the right (KI 19), in the left-hand margin of page 37.

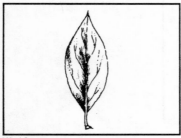

KI 18

3. If your specimen is a vine with either simple or compound leaves, find the picture to the right (KI 204), in the left-hand margin of page 407.

35A

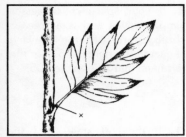

KI 53

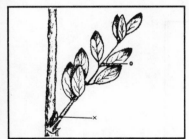

KI 19

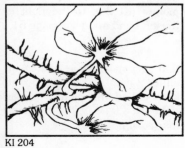

KI 204

1. If the leaf stalks (not the bladelet stalks) are opposite each other on the stem, proceed to the picture, shown to the right (KI 20), in the left-hand margin of page 39.

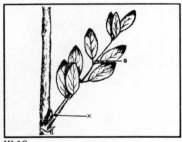

KI 19

2. If the leaf stalks (again do not look at the bladelet stalks) are not opposite each other, find the picture (KI 26), to the right in the left-hand margin of page 51.

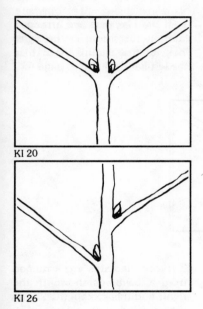

KI 20

KI 26

1. If the bladelets are arranged laterally, as shown to the right, look for this picture (KI 21), in the left-hand margin of page 41.

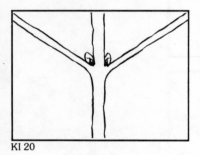

KI 20

2. If the bladelets are arranged from a point, as shown to the right, find this symbol (KI 23), in the left-hand margin of page 45.

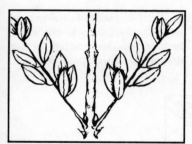

KI 21

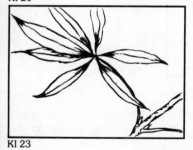

KI 23

1. Look at the edge of the bladelet, if it is smooth (entire), it is TEXAS PORLIERIA.

2. If the bladelet margin is single-toothed like a sawblade, as shown to the right, go to this symbol (KI 22), on page 43.

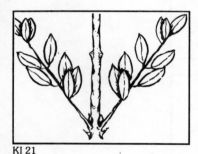

KI 21

3. If the bladelet margin is double-toothed, it is YEL-LOW-ROOT.

Fig. 29

TEXAS PORLIERIA
(Porlieria angustifolia)
A shrub or small tree up to 20 ft. high with stiff, short branches. The purple flowers are fragrant, velvety, have 10 petals and 10 stamens. The heart-shaped fruit with 3 terminal prongs is distinctive. This shrub occurs in central, western and southwestern Texas in our area. It belongs to the caltrop family *(Zygophyllaceae)*.

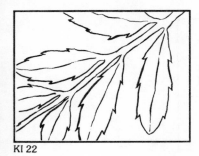
KI 22

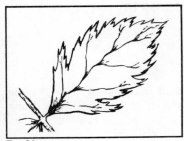
Fig. 30

YELLOW-ROOT
(Xanthorhiza simplicissima)
A weak shrub up to 2 ft. high, growing in moist, shady forests of Louisiana and southeast Texas. It has small, purplish flowers with 5 petaloid sepals and 5-10 stamens. It belongs to the buttercup family *(Ranunculaceae)*.

42

1. If the leaflet looks like the picture to the right and has almost no leaflet stalk, it is BOX-ELDER.

2. If the leaflet looks like the picture to the right and the young twigs are gray, it is an ASH.

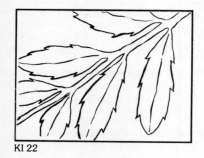

KI 22

3. If the leaflet looks like the picture to the right with its tip being very slender, it is YELLOW TRUMPET.

4. If the bladelet looks like the picture to the right and its young twigs are grooved, it is ELDERBERRY.

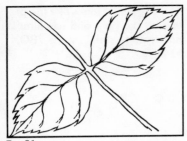

Fig. 31

BOXELDER
(Acer negundo)
A small tree, reaching at most 75 ft. in height; of poor growth form, the crown rounded and the young stems green. The flowers are inconspicuous; the fruit being a double, greenish samara. It grows among streams in all of our area except the western half of Texas and New Mexico. It belongs to the maple family *(Aceraceae)*.

Fig. 32

ASH
(Fraxinus spp.)
A complex of trees growing up to 70 ft. high or more and occurring along streams in all states of our area. Their minute flowers are inconspicuous but their single-bladed samaras are distinctive. They belong to the olive family *(Oleaceae)*.

Fig. 33

YELLOW TRUMPET
(Tecoma stans)
A small shrub up to 3 ft. high with bright yellow, trumpet-shaped flowers 3-5 in. long. The fruit is a long, bean-shaped capsule 4-6 in. long. It grows on rocky, open slopes in Trans-Pecos Texas and New Mexico. It belongs to the bignonia family *(Bignoniaceae)*.

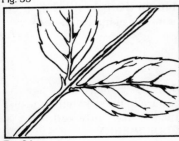

Fig. 34

ELDERBERRY
(Sambucus spp.)
Shrubs with many stems, growing in muddy areas and along streams, usually 3 to 12 ft. high. The flowers are yellowish-white in terminal flat-topped clusters. The fruits are dark berry-like drupes about ¼ in. in diameter. They belong to the honeysuckle family *(Caprifoliaceae)*.

1. Count the bladelets, if there are 5 or more, as shown in (KI 24) to the right, go to (KI 24), in the left-hand margin of page 47.

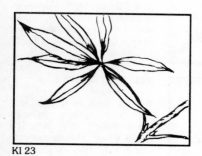

KI 23

2. If you find less than 5 leaflets (some leaves may have as many as five but no more), find the picture to the right (KI 25), in the left-hand margin of page 49.

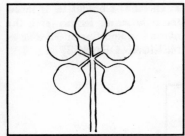

KI 24

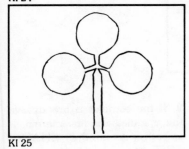

KI 25

1. If the bladelet is narrow throughout its length and the margin is knotty-glandular, it is MEXICAN ORANGE.

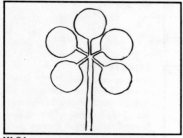

KI 24

2. If the bladelet is broadest in the middle and its margin is smooth, it is the CHASTE TREE.

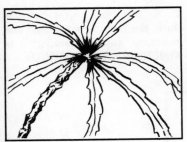

Fig. 35

MEXICAN ORANGE
(Choisya dumosa)
A rounded shrub 3-6 ft. high with showy, white flowers. Uncommon in the mountains of Trans-Pecos Texas and New Mexico. It belongs to the citrus family *(Rutaceae).*

Fig. 36

CHASTE-TREE
(Vitex agnus-castus)
A many-stemmed shrub up to 15 ft. high with aromatic foliage and blue 2-lipped flowers. Introduced from China, this species has escaped in various parts of our area, mostly sandy soils. It is a member of the verbena family *(Verbenaceae).*

1. If the leaflets are 3 and the bladelet margin is smooth, it is the HOP TREE.

2. If the leaflets are 3 and the margin sawtoothed, it is AMERICAN BLADDERNUT.

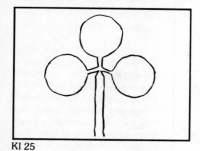

KI 25

3. If the leaflets are 4-5 with a sawtoothed margin, it is a BUCKEYE.

Fig. 37

HOP TREE
(Ptelea trifoliata)
Mostly a rounded shrub with ill-scented foliage, small greenish-white flowers and winged fruits. It grows in a wide range of soils in all of our area. It belongs to the citrus family *(Rutaceae)*.

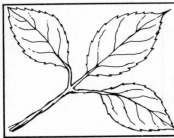
Fig. 38

AMERICAN BLADDERNUT
(Staphylea trifolia)
A shrub or small tree up to 20 ft. high with striped twigs, small greenish-white flowers and bladder-like fruits 2 in. long. It grows in rich soil in shaded forests in Oklahoma and Arkansas. It belongs to the bladdernut family *(Staphyleaceae)*.

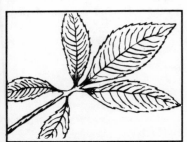
Fig. 39

BUCKEYE
(Aesculus spp.)
Mostly shrubs with showy yellow, red or yellowish-green tubular flowers and globose vestured fruits. They grow mostly along streams over most of our area. They belong to the buckeye family *(Hippocastanaceae)*.

50

1. If the bladelets are arranged laterally, as shown to the right in (KI 30), look for this picture in the left-hand margin of page 59.

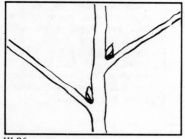

KI 26

2. If the bladelets are arranged from a point, as shown to the right (KI 27), look for this picture in the left-hand margin of page 53.

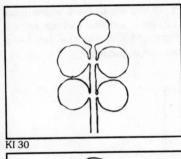

KI 30

KI 27

52

1. If the leaflets are more than 20, look for the picture to the right (KI 28), in the left-hand margin of page 55.

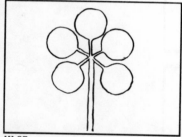

KI 27

2. If the leaflets are 3, proceed to the picture, shown to the right (KI 29), in the left-hand margin of page 57 and 57A. *Caution:* Your specimen may be poisonous to the touch.

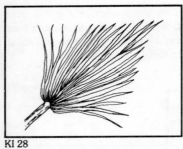

KI 28

KI 29

1. If this tree lacks a trunk and the leaves arise from a crown at ground level, it is DWARF PALM.

2. If this tall tree has a trunk and the bladelets are smooth, it is LOUISIANA PALM.

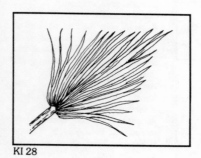

KI 28

3. If this low tree has a trunk and the bladelets are saw-toothed, it is SAW PALMETTO.

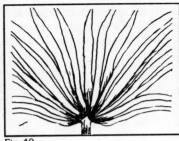

Fig. 40

DWARF PALM
(Sabal minor)

A woody plant with an underground stem, the fanlike leaves arising from ground level. Small flowers are in clusters subtended by long bracts and the fruits are round, black and ⅓ in. in diameter. This palm is abundant in stream bottoms and swamps of southern Louisiana and southeastern Texas. It belongs to the palm family *(Palmae)*.

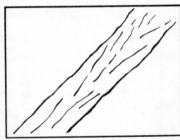

Fig. 41

LOUISIANA PALM
(Sabal louisiana)

A small tree 3-6 ft. tall with a basal zone of adventitious roots and leaves crowded toward the top of the trunk. Tiny white flowers are subtended by long-pointed spathes (bracts), the fruits brown to black and sphaerical and about ¼ in. in diameter. It occurs uncommonly in southern Louisiana. It belongs to the palm family *(Palmae)*.

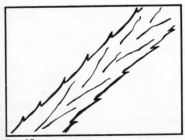

Fig. 42

SAW PALMETTO
(Serenoa repens)

A decumbent or erect woody plant with clustered erect or ascending fan-shaped leaves. The very small flowers are bracted in clusters shorter than the leaves. The ovoid fruit is black and about ¾ in. long. It occurs in sandy habitats in eastern Louisiana. It belongs to the palm family *(Palmae)*.

56

1. If the bladelet is lobed, it is POISON OAK.

2. If the bladelet is mostly entire, it is POISON IVY.

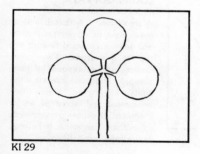

KI 29

3. If the bladelet is cut at the apex, it is SKUNK BUSH.

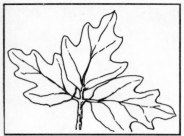

Fig. 43

POISON OAK
(Rhus toxicodendron var. quercifolia)
Differs from Poison Ivy only in the lobing of its leaflets.

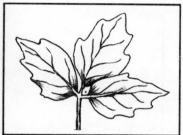

Fig. 44

POISON IVY
(Rhus toxicodendron var. *vulgaris)*
Low shrub or vine with poisonous, toothed foliage. The flowers are small, greenish-white and the waxy fruits are dull-white, globose and about ¼ in. in diameter. It occurs in forests or stream bottoms throughout our area. It belongs to the sumac family *(Anacardiaceae)*.

Fig. 45

SKUNK BUSH
(Rhus aromatica var. flabelliformis)
An ill-scented shrub up to 6 ft. high. The flowers are small, greenish-white and hairy. The reddish fruit is round and ¼ in. in diameter. It occurs mostly in limestone soils in the western half of Texas and New Mexico. It belongs to the sumac family *(Anacardiaceae)*.

4. If two bladelets are smaller than the third and the margin is toothed, it is FRAGRANT SUMAC.

KI 29

5. If two bladelets are smaller than the third but their margins are smooth, it is HOP TREE.

57A

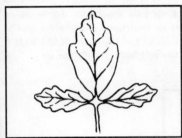

Fig. 46

FRAGRANT SUMAC
(Rhus aromatica)
A pungently-fragrant shrub up to 6 ft. high, usually growing in thickets. Flowers hairy, small and yellowish-green. The fruits are rounded, red, hairy and ⅓ in. in diameter. It occurs in the forests, on rocky outcrops and dunes of eastern Texas, Oklahoma, Arkansas and Louisiana. It belongs to the sumac family *(Anacardiaceae)*.

Fig. 47

HOP TREE
(Ptelea trifoliata)
Mostly a rounded shrub with ill-scented foliage, small greenish-white flowers and winged fruits. It grows in a wide range of soils in all of our area. It belongs to the citrus family *(Rutaceae)*.

58A

1. If your specimen possesses spines or thorns on its stems or leaf stalks, as illustrated in (KI 31), to the right, go to (KI 31), in the left-hand margin of page 61.

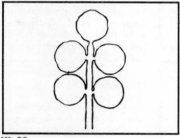

KI 30

2. If your specimen is neither prickly nor thorny, or prickly only on the bladelet, go to the picture, at the right (KI 39), in the left-hand margin of page 77.

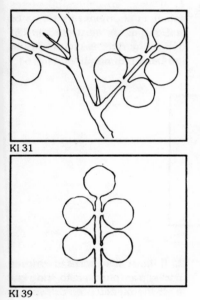

KI 31

KI 39

1. If the spines-thorns are straight, as shown to the right in (KI 32), go to this picture in the left-hand margin of page 63.

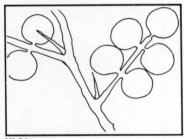

KI 31

2. If the spines-thorns are mostly curved, as pictured to the right in (KI 36), find this drawing in the left-hand margin of page 71.

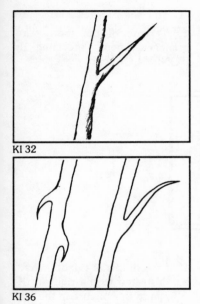

KI 32

KI 36

1. If some of the thorns are branched, your specimen is HONEY LOCUST.

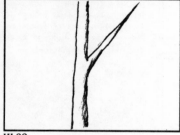

KI 32

2. If the thorns are unbranched, as illustrated to the right in (KI 33), go to this picture in the left-hand margin of page 65.

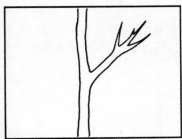

Fig. 48

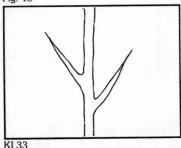

KI 33

HONEY LOCUST
(Gleditsia tricanthos)

A loose-crowned tree up to 100 ft. tall. Flowers mostly imperfect, slightly irregular, ¼ in. long. The fruit is a flattened pod ½-1½ ft. long and 1-1½ in. wide, dark-brown and shiny when mature. It grows in moist, rich soil throughout our area except for the western part of Texas and New Mexico. It belongs to the pea family *(Leguminosae)*.

1. If the leaflet is 1 to 1½ in. or less long, it is HONEY MESQUITE.

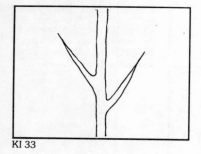

KI 33

2. If the leaflet is ⅛-⅓ in. long, find the drawing (KI 34), shown to the right and in the left-hand margin of page 67.

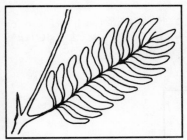

Fig. 49

HONEY MESQUITE
(Prosopis glandulosa)
Shrubs to medium-sized trees with drooping, crooked branches. Flowers yellowish-green, small; fruit a pod that is narrow and 4-9 in. long. Often abundant in disturbed grasslands throughout our area. It belongs to the pea family *(Leguminosae)*.

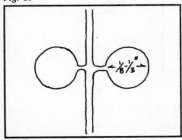

KI 34

1. If the leaflets are 4-6, it is PALO VERDE.

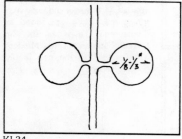

KI 34

2. If the leaflets are 7 to many, go to (KI 35), shown to the right and in the left-hand margin of page 69.

Fig. 50

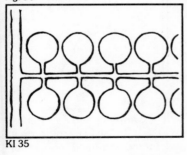

KI 35

PALO VERDE
(Cercidium spp.)
Green-twigged shrubs or small trees, many-branched and up to 25 ft. tall. Flowers are yellow and ½ in. long. The fruit is a pod 1-2½ in. long, about ½ in. wide and dark brown at maturity. The two species, *C. macrum* and *texanum,* intergrade. They are found in Trans-Pecos Texas, on the Rio Grande Plain and in New Mexico. They belong to the pea family *(Leguminosae).*

1. If the thorns are long and occur together, mostly in pairs, it is an ACACIA.

2. If the thorns are short and mostly single and the leaflet very narrow, it is JERUSALEM THORN.

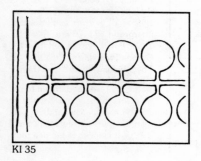

KI 35

3. If the short thorns are single and the leaflet relatively broad, it is TEXAS EBONY.

Fig. 51

ACACIA
(Acacia spp.)
A complex of spiny shrubs and small trees with very small white to yellow flowers, each containing 20 to 100 stamens. The fruit is an elongate pod. They are common plants of desert flats and hills in New Mexico, far western and southwestern Texas. They belong to the legume family *(Leguminosae)*.

Fig. 52

JERUSALEM THORN
(Parkinsonia aculeata)
Shrubs or medium-sized trees with green bark, slender-spreading branches and a rounded crown. Showy, yellow flowers become red-dotted with age. The fruit is an elongate constricted pod 2-4 in. long and brown to red in color. It is found usually in moist, sandy soil and in the southern part of New Mexico and Texas. It belongs to the pea family *(Leguminosae)*.

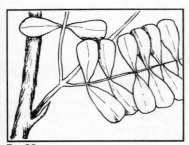

Fig. 53

TEXAS EBONY
(Pithecellobium flexicaule)
Shrubs but mostly trees up to 45 ft. high with dense green foliage and short forked or zig-zag branches. The minute flowers are yellowish to whitish in dense, short spikes. The fruit is a woody pod 4-6 in. long and 1 in. wide and dark brown to black in color. It is a legume *(Leguminosae)*.

1. If the leaf stalk has wings at its base, it is a ROSE.

2. If the leaf stalk outgrowths appear as illustrated to the right, it is an ACACIA.

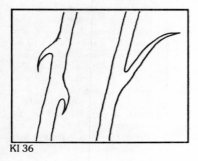

KI 36

3. If the leaf stalk has no outgrowths at its base, find the picture (KI 37), to the right and in the left-hand margin of page 73 and 73A.

Fig. 54

ROSE
(Rosa spp.)
Shrubs with recurved stem prickles. The flowers are typically showy with 5 petals and many stamens, ranging in color from yellow to pink to white. The fruit is an enlarged floral cup or "hip." The group is a complex of often difficult to separate species and hybrids occurring throughout our area. They belong to the rose family *(Rosaceae).*

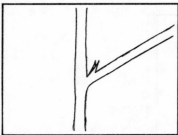

Fig. 55

ACACIA
(Acacia spp.)
A complex of spiny shrubs and small trees with very small white to yellow flowers, each containing 20 to 100 stamens. The fruit is an elongate pod. They are common plants of desert flats and hills in New Mexico, far western and southwestern Texas. They belong to the legume family *(Leguminosae).*

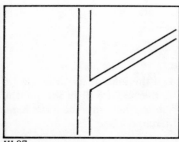

KI 37

72

1. If the leaflet is triangular in outline and the leaflets only 3, it is an ERYTHRINA.

2. If the leaflets are 3-7, it is a BRAMBLE.

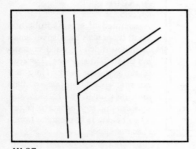

KI 37

3. If the leaflets are 5 to 13, go to the picture (KI 38), to the right, and in the left-hand margin of page 75.

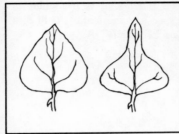

Fig. 56

ERYTHRINA
(Erythrina spp.)
Shrubs or sometimes small trees up to 15 ft. high with elongate, showy red flowers. The fruit is an elongate, constricted pod. WESTERN ERYTHRINA grows in the desert mountains of southwestern New Mexico and EASTERN ERYTHRINA grows in sandy soils of eastern Texas and Louisiana. They are legumes *(Leguminosae).*

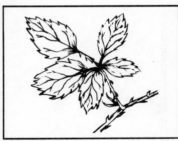

Fig. 57

BRAMBLE
(Rubus spp.)
Prostrate to erect shrubs with arching, prickled stems. Flowers are mostly white with 5 petals and many stamens. A large, taxonomically difficult complex of mostly hybrids variously also called blackberries or dewberries; the fruit being a cluster of dark-colored druplets. Many grow in damp sands but they are also found in dry habitats throughout our area. They belong to the rose family *(Rosaceae).*

KI 38

74

4. If the leaflets are 13 to many and their width ½ to 1 in., it is ROBINIA.

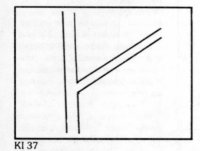

KI 37

5. If the leaflets are many and only ½ in. or less wide, it is a MIMOSA.

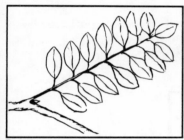

Fig. 58

ROBINIA
(Robinia spp.)

A complex of small, spiny trees and shrubs that have similar characteristics. The pea flowers are white to rose color, showy and quite hairy. The fruit is an elongate pod 2-3 in. long. BLACK LOCUST and ROSE-ACACIA occur in Arkansas and Oklahoma; whereas the RUSBY and NEW MEXICAN LO-CUSTare found in the mountains of New Mexico.They belong to the pea family *(Leguminosae).*

Fig. 59

MIMOSA
(Mimosa spp.)

Armed, mostly low shrubs up to 8 ft. high. The minute flowers are borne in heads that look like fluffy balls because of the many exserted stamens. The fruit is an elongate, often prickly pod. Mostly desert plants of southwestern Texas and southern New Mexico. They belong to the pea family *(Leguminosae).*

1. If the leaflet margin is smooth, it is PRICKLY-ASH.

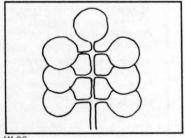

KI 38

2. If the leaflet margin is serrate, it is the DEVIL'S WALKING STICK.

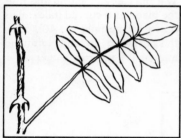
Fig. 60

PRICKLY-ASH
(Zanthoxylum spp.)
Armed shrubs or trees up to 25 ft. high. Conspicuous tubercles on the trunk are distinctive. The flowers are small, yellowish-green and the fruits are inconspicuous follicles. They occur in open forests in all of the states of our area except for New Mexico. They belong to the citrus family *(Rutaceae).*

Fig. 61

DEVIL'S WALKING STICK
(Aralia spinosa)
A shrub or tree up to 35 ft. high with few branches and leaves 3-4 ft. long. The flowers are small, ⅛ in. wide, but are clustered into showy white panicles. The fruit is a black drupe ¼ in. in diameter and 3-5 angled. It grows in rich moist soil in Arkansas, eastern Oklahoma, eastern Texas and Louisiana. It belongs to the ginseng family *(Araliaceae).*

1. If the bladelet midrib and thus the bladelet is curved (falcate) as drawn to the right (KI 40), go to this drawing in the left-hand margin of page 79.

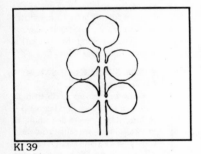

KI 39

2. If the bladelet midrib is straight and thus the bladelet not curved sideways, find the drawing (KI 41) to the right in the left-hand margin of page 81.

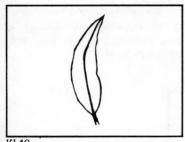

KI 40

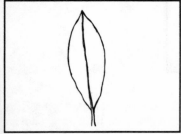

KI 41

1. If the bladelet margin is saw-toothed, it is the PECAN.

2. If the bladelet margin is smooth and ⅓ to 1 in. long, it is PISTACIO.

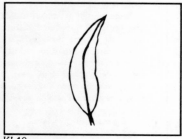

KI 40

3. If the bladelet margin is smooth and 1½-4 in. long, it is SOAPBERRY.

79

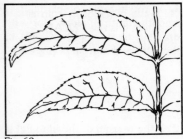

Fig. 62

PECAN
(Carya illinoensis)
The Illinois hickory is a handsome tree up to 150 ft. tall with a broad-rounded crown. The flowers are inconspicuous and borne in separate catkins on the same tree. The fruit is a brownish endocarp 1-3 in. long, splitting out of a 4-valved "husk." It grows along floodplains in Texas, Arkansas, Oklahoma and Louisiana. It belongs in the walnut family *(Juglandaceae)*.

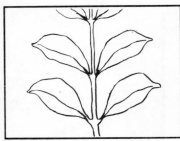

Fig. 63

PISTACIO
(Pistacia texana)
A large shrub or multi-trunked tree up to 12 ft. tall with distinctive red spring foliage. Tiny flowers are grouped in reddish-yellow clusters. The fruit is oval, reddish-brown and ¼ in. long. It is found in limestone stream beds in southwestern Texas. It belongs to the sumac family *(Anacardiaceae)*.

Fig. 64

SOAPBERRY
(Sapindus saponaria var. *drummondii)*
Trees up to 50 ft. high with grayish bark and erect branches. The tiny white flowers are borne in showy panicles and the distinctive fruit is translucent, yellowish, wrinkled and about ½ in. in diameter. It grows along streams and fencerows in all of the states of our area. It belongs to the soapberry family *(Sapindaceae)*.

80

1. If the margin of bladelet is smooth (entire), look for the drawing to the right (KI 44) in the left-hand margin of page 87.

2. If the margin of the bladelet is toothed and the teeth not spiny-tipped, as drawn to the right, find this picture (KI 42) in the left-hand margin of page 83 and 83A.

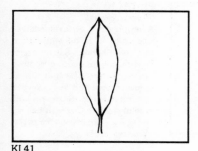

KI 41

3. If the margin of the bladelet is toothed and the teeth are spiny, it is a MAHONIA.

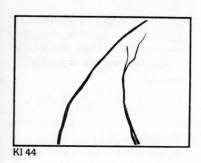
KI 44

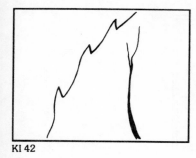
KI 42

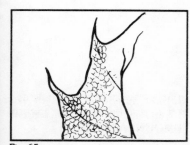
Fig. 65

MAHONIA
(Berberis spp.)
A seven-species complex with similar characteristics, consisting of shrubs up to 8 ft. high identified by their holly-like, spiny foliage. The flowers are small, yellow, and the fruits are round, reddish, pulpy berries. They grow in the dry mountains of west Texas and New Mexico. They belong to the barberry family *(Berberidaceae)*.

1. Count the number of bladelets, if there are mostly more than 9, look for the diagram to the right (KI 43) on page 85.

2. If the number of bladelets are 3-5, it is SHAGBARK HICKORY.

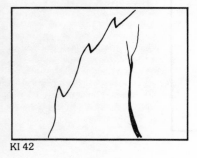

KI 42

3. If the number of bladelets are 3-7 and each is ½ to 1 in. long, it is MOUNTAIN ASH.

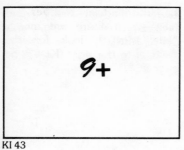

KI 43

Fig. 66

SHAGBARK HICKORY
(Carya ovata)
A tree up to 100 ft. tall with an oblong crown and shaggy bark. Flowers inconspicuous, the fruit a 4-angled, ovoid, pale endocarp. It is found in rich bottomlands in northeastern Texas, eastern Oklahoma, Louisiana and Arkansas. It belongs to the walnut family *(Juglandaceae)*.

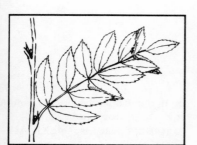

Fig. 67

MOUNTAIN ASH
(Sorbus dumosa or *scopulina)*
Shrubs up to 15 ft. high with slender branches, the two species differing slightly in their floral clusters; flowers small, white; fruits ¼ in. in diameter, round and orangish. They grow in the upper montane forest of New Mexico and belong to the rose family *(Rosaceae)*.

84

4. If the bladelets are 7-9 and each is 1½-4 in. long, it is CHINABERRY.

5. If the bladelets are 5-9 and the bladelet shape is oval (ovate), it is SHELLBARK HICKORY.

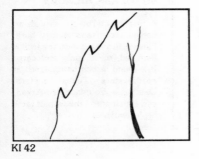

KI 42

6. If the bladelets are 5-7 and the bladelet lance-shaped (lanceolate), it is MEXICAN BUCKEYE.

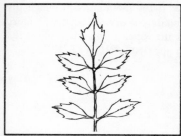

Fig. 68

CHINABERRY
(Melia azedarch)
A weak-wooded tree up to 45 ft. high with rounded crown and twice-compound leaves often 2 ft. long. Flowers ½ in. wide, purplish; fruit roundish, ½-¾ in. in diameter, yellow, fleshy and translucent. It is an Asiatic plant that has escaped and now grows in Texas, Oklahoma, Arkansas and Louisiana. It belongs to the mahogany family *(Meliaceae)*.

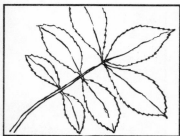

Fig. 69

SHELLBARK HICKORY
(Carya laciniosa)
A narrow-crowned short-branched tree up to 120 ft. tall. Flowers inconspicuous, the fruit a thick-shelled, tannish endocarp. It grows on rich floodplains in Louisiana, northeast Texas, Oklahoma and Arkansas. It belongs to the walnut family *(Juglandaceae)*.

Fig. 70

MEXICAN BUCKEYE
(Ungnadia speciosa)
A shrub or tree up to 30 ft. high with small upright branches and an irregular crown. The irregular flowers are showy, fragrant and rose-colored. The pear-shaped fruit is rough, leathery, reddish-brown and 2 in. broad. It grows in limestone stream beds from central Texas south and west into New Mexico. It belongs to the soapberry family *(Sapindaceae)*.

84A

1. If the main stalk to which the bladelets are attached is hairy with sticky glands (like flypaper), it is BUTTERNUT.

2. If the main stalk to which the bladelets are attached is smooth or slightly hairy and the bladelets are 11-23 in number, it is BLACK WALNUT.

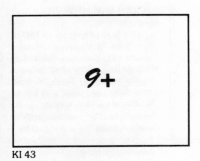

KI 43

3. If the bladelets and stalks are non-hairy and the bladelets number 11-31, it is SMOOTH SUMAC.

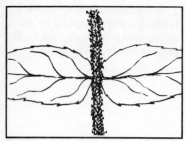
Fig. 71

BUTTERNUT
(Juglans cinerea)
Short-trunked tree up to 100 ft. high with a round-topped crown, gray diamond-patterned outer bark and green inner bark. The flowers are inconspicuous and imperfect and the drupe is 2-3 in. long, densely matted with red hairs, thick-hulled with a light brown nut. It grows scattered in mixed forests on rich soils in northeastern Oklahoma and Arkansas. It belongs to the walnut family *(Juglandaceae)*.

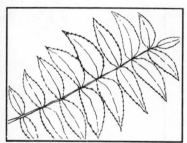
Fig. 72

BLACK WALNUT
(Juglans nigra)
A dark-barked tree up to 150 ft. high with a massive rounded crown. Flowers are imperfect and inconspicuous. The drupe is almost round, aromatic, indehiscent, yellowish-green and 1-2 in. in diameter. It grows along streams and in rich woods in all states of our area except New Mexico. It belongs to the walnut family *(Juglandaceae)*.

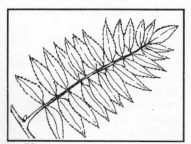
Fig. 73

SMOOTH SUMAC
(Rhus glabra)
Shrub or small tree up to 10 ft. high with glaucous branches. Flowers in terminal clusters, white, very small; the fruit small and covered with red, velvety hairs, giving the terminal cluster a flaming appearance in the fall. It grows in rich soil in all states in our area. It belongs to the sumac family *(Anacardiaceae)*.

86

1. If the leaflets range in number from 3 to 11 per leaf, go to drawing (KI 45), as shown to the right, in the left-hand margin of page 89 and 89A.

2. If the leaflets range in number from 11 to 21 on the average per leaf, find the diagram to the right (KI 46), in the left-hand margin of page 91.

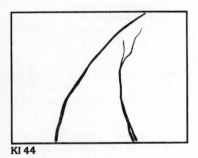

KI 44

3. If the leaflets range in number from 20 to 30 on the average per leaf, find the illustration to the right (KI 50), in the left-hand margin of page 99.

4. If the leaflets' count is greater than 32 for any of the leaves (some leaves may be as low as 11), go to the symbol, shown to the right (KI 51), in the left-hand margin of page 101.

3-11

KI 45

11-21

KI 46

20-30

KI 50

32+

KI 51

1. If the bladelets are 5-11 and each only 1/25-1/6 in. long, it is DALEA.

2. If the bladelets are 5-9 and each may be ¾ -2 in. long, it is EVERGREEN SUMAC.

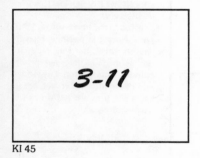

KI 45

3. If the bladelets are 3-7 and each may be ½-1 in. long, it is SHRUBBY CINQUEFOIL.

Fig. 74

DALEA
(Dalea spp.)
Western low shrubs with smooth slender stems. Flowers variously colored, pea-shaped, the fruit a short pod. They grow in dry soil in Texas and New Mexico and belong to the pea family *(Leguminosae)*.

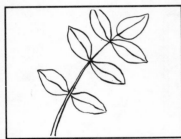

Fig. 75

EVERGREEN SUMAC
(Rhus virens)
Shrub or small tree up to 12 ft. high growing in patches. Flowers whitish, small, the fruit red-hairy and ¼ in. long. It grows on dry hillsides from central Texas west into New Mexico. It belongs to the sumac family *(Anacardiaceae)*.

Fig. 76

SHRUBBY CINQUEFOIL
(Potentilla fruiticosa)
Low shrub up to 4 ft. high, often sprawling. Flowers large, showy and yellow, the fruit a hairy achene. An alpine plant found above timberline in boggy meadows in New Mexico. It belongs to the rose family *(Rosaceae)*.

4. If the main stalk between the leaflets is winged, it is LITTLE-LEAF SUMAC.

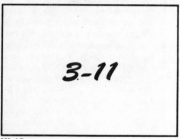

KI 45

5. If the bladelets are 4-6 and each is ⅛-¼ in. long, it is WISLIZENUS SENNA.

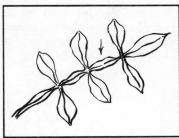
Fig. 77

LITTLE-LEAF SUMAC
(Rhus microphylla)
A clump-forming, highly-branched shrub, growing up to 15 ft. high. The tiny flowers are greenish-white in stiff spikes; the fruit is ovoid, reddish-orange and about ¼ in. long. It grows on mesas or rocky hillsides in Texas and New Mexico. It belongs to the sumac family *(Anacardiaceae)*.

WISLIZENUS SENNA
(Cassia wislizenii)
A shrub up to 10 ft. tall with many short branches. The flowers are yellow and ¾-1½ in. wide. The fruit is a slender pod 3-6 in. long. An escape from Argentina, it grows in rocky soils of western Texas and New Mexico. It belongs to the pea family *(Leguminosae)*.

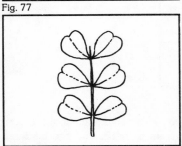
Fig. 78

1. Count the bladelets of several leaves, if some of them exceed 15, look for the drawing at the right (KI 47) in the left-hand margin of page 93.

2. If the bladelet count ranges from 7 to 13, and the bladelets are 2½-4 in. long, it is the dangerous POISON SUMAC.

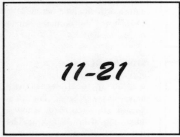

KI 46

3. If the bladelet count ranges from 5 to 13, and the bladelets are 1-2½ in. long, it is the MESCAL-BEAN.

4. If the bladelets are 4-6 and each is ⅔ in. long, it is TEXAS PORLIERIA.

KI 47

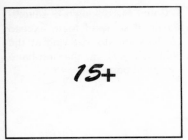

Fig. 79

Fig. 80

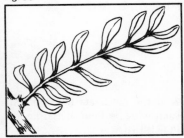

Fig. 81

POISON SUMAC
(Rhus vernix)
Shrub or small tree up to 20 ft. high with glaucous and later gray twigs. Flowers tiny and green. The fruit is round, whitish and up to ¼ in. in diameter. It grows in swamps or boggy ground in eastern Texas and Louisiana. It belongs to the sumac family *(Anacardiaceae)*.

MESCAL-BEAN
(Sophora secundiflora)
A shrub up to 30 ft. tall with thick, glossy foliage. The pea-shaped flowers are showy, violet-colored and ¾ in. long. The fruit is an elongate, constricted pod 1-5 in. long. It grows in limestone soils from central Texas west into New Mexico. It belongs to the pea family *(Leguminosae)*.

TEXAS PORLIERIA
(Porlieria angustifolia)
A shrub or small tree up to 20 ft. high with stiff, short branches. The purple flowers are fragrant, velvety, with 10 petals and 10 stamens. The heart-shaped fruit with 3 terminal prongs is distinctive. This shrub occurs in central, western and southwestern Texas in our area. It belongs to the caltrop family *(Zygophyllaceae)*.

92

1. If the bladelet stalk is winged, it is WINGED SUMAC.

2. If the bladelet stalk is hairy, it is an INDIGO-BUSH.

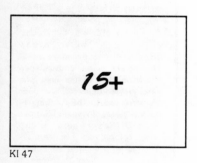

KI 47

3. If the bladelet stalk is smooth, go to the symbol at the right (KI 48) in the left-hand margin of page 95.

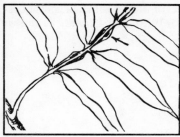
Fig. 82

WINGED SUMAC
(Rhus copallina)
A large shrub or small tree up to 30 ft. high with small greenish-white flowers and flattened red-hairy fruits ⅛ in. in diameter. It grows in bottomlands and on rocky hills in Louisiana and Arkansas and the eastern halves of Texas and Oklahoma. It belongs to the sumac family, *(Anacardiaceae)*.

Fig. 83

INDIGO-BUSH
(Amorpha spp.)
A complex of low to tall shrubs, growing in moist areas throughout our five-state area. The flower is an unusual pea flower in that it has only one petal instead of five. Their pods are very short. They belong to the pea family *(Leguminosae)*.

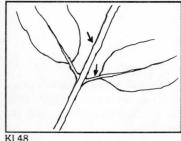
KI 48

94

1. If the bladelet length is ⅛-3/16 in. and appears as drawn to the right, it is POINCIANA.

2. If some of the leaflets are as much as 2-2½ in. long, it is an INDIGO-BUSH.

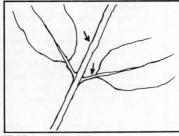

KI 48

3. If some of the leaflets are as much as 1-1½ in. in length, find (KI 49), as shown to the right, on page 97.

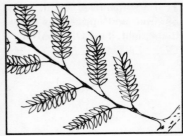

Fig. 84

POINCIANA
(Caesalpinia gilliesii)
A shrub or small tree up to 15 ft. high with green, brown or reddish branches. The showy flowers are yellow with long, exserted, red, upcurved stamens. The fruit is a flat pod, 1½ in. long. Both fruit and flower stalks are covered with red, stalked glands. A native of Argentina, it grows in dry habitats in central and western Texas and New Mexico as an escape from cultivation. It belongs to the pea family *(Leguminosae)*.

Fig. 85

INDIGO-BUSH
(Amorpha spp.)
A complex of low to tall shrubs, growing in moist areas throughout our five-state area. The flower is an unusual pea flower in that it has only one petal instead of five. Their pods are very short. They belong to the pea family *(Leguminosae)*.

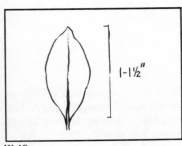

1-1½"

KI 49

1. If the leaf stalk has outgrowths from its base, as depicted to the right, it is GOLDEN-BALL LEAD-TREE.

2. If stalk outgrowths are absent and the leaflet tips pointed, it is EVE'S NECKLACE.

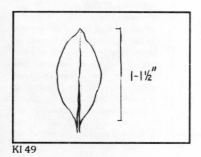

KI 49

1-1½"

3. If stalk outgrowths are absent and the leaflet tip is blunt or notched, it is SMOOTH INDIGO.

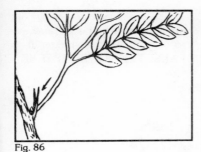
Fig. 86

GOLDEN-BALL LEAD-TREE
(Leucaena retusa)
A shrub or small tree 6-20 ft. tall with brittle stems. The tiny flowers are borne in distinctive, globose heads about 1 in. in diameter, appearing as fuzzy, yellow balls. The fruits are long (3-10 in.) narrow pods. It grows on dry limestone slopes in central and Trans-Pecos Texas west into New Mexico. It belongs to the pea family *(Leguminosae)*.

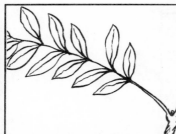
Fig. 87

EVE'S NECKLACE
(Sophora affinis)
A shrub or small tree up to 30 ft. high with spreading branches. Its pea-shaped flowers are about ½ in. long and are white to pink. The fruit is a distinctive black constricted legume up to 4 in. long and looks like a string of beads. It grows in limestone soils in northwestern Louisiana and southwestern Oklahoma, ranging down through central Texas. It is a member of the pea family *(Leguminosae)*.

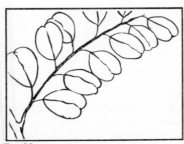
Fig. 88

SMOOTH INDIGO
(Amorpha texana or *laevigata)*
These hairless shrubs are 3-9 ft. tall and are considered by many taxonomists as the same species. See, also, notes for INDIGO-BUSH (pg. 96). *A. texana* grows along streams in limestone areas of central and southwestern Texas. *A. laevigata* is found in moist, rich soil in northeastern Texas, Oklahoma, Louisiana and Arkansas. They belong to the pea family *(Leguminosae)*.

1. If the bladelets range in number from 12 to 24 and are only 1/12-⅛ in. long, it is FALSE MESQUITE.

2. If the bladelets range in number from 15 to 31 and are 1/5 to ⅜ in. long, it is TEXAS KIDNEYWOOD.

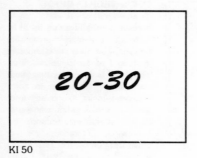

KI 50

3. If the bladelets number from 9 to 27 and some are as long as 2 in., it is an INDIGO-BUSH.

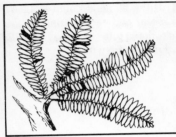

Fig. 89

FALSE MESQUITE
(Calliandra conferta or *eriophylla)*
These two species are easily con-
fused by taxonomists. They are
low, densely-branched shrubs up
to 3 ft. tall. Their small white to
purple flowers are clustered into
heads, prominent by the many
long stamens. The fruit is a narrow pod 1-3 in. long. *Callindra
conferta* grows in limestone soils
in southwestern Texas and *C.
eriophylla* edges into Trans-Pecos
Texas from New Mexico. They
belong to the pea family
(Leguminosae).

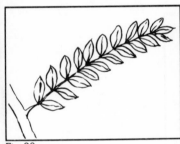

Fig. 90

TEXAS KIDNEYWOOD
(Eysenhardtia texana) An
irregularly-shaped shrub up to 9
ft. tall with glandular, aromatic
foliage. The pea flowers are
small and white to yellow. The
fruit is a pod ½ in. long and
gland-dotted. It grows on dry
calcareous hills in central and
southwestern Texas. It belongs to
the pea family *(Leguminosae).*

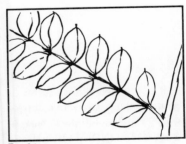

Fig. 91

INDIGO-BUSH
(Amorpha spp.)
A complex of low to tall shrubs,
growing in moist areas
throughout our five-state area.
The flower is an unusual pea
flower in that it has only one
petal instead of five. Their pods
are very short. They belong to
the pea family *(Leguminosae).*

100

1. If the bladelet count is 11-41 per leaf and each bladelet is 2-5 in. long, it is the TREE-OF-HEAVEN.

2. If the bladelet count is 11-45 per leaf and the leaflets are extremely narrow, 1/12 in. or less, it is the DUNE BROOM.

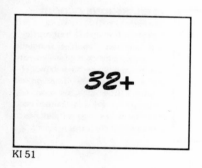

32+

KI 51

3. If the bladelets are shorter or broader than the above, go to (KI 52) in the left-hand side of page 103.

Fig. 92

TREE-OF-HEAVEN
(Ailanthus altissima)
A graceful tree up to 60 ft. tall with smooth bark and a globose crown. Its small greenish-yellow flowers are borne in loose clusters. The fruit is a paired samara with brownish-red wings that are notched on one side. It is native to Australasia, escaping and growing in waste places in the eastern third of Texas, Oklahoma, Arkansas and Louisiana. It belongs to the quassia family *(Simarubaceae)*.

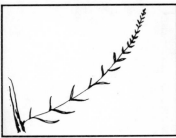

Fig. 93

DUNE BROOM
(Parryella filifolia)
Low many-branched shrubs up to 3 ft. tall with glandular foliage. Its yellowish-green pea flowers are ½ in. long and are borne in lax, terminal clusters. The fruit is a glandular pod ¼ in. long. It grows on sandy hillsides in New Mexico. It belongs to the pea family *(Leguminosae)*.

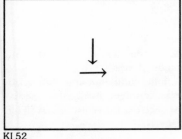

KI 52

1. If the stems are deeply grooved and the bladelet count is 40-60 per leaf, it is PRAIRIE ACACIA.

2. If the stems are neither grooved nor angled and the bladelets are densely gray-hairy, it is LEADEN INDIGO-BUSH.

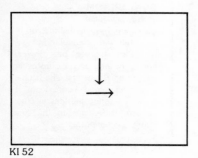

KI 52

3. If the stems and bladelets are smooth, and the bladelet stalk is grooved, it is KIDNEYWOOD.

4. If the stems, bladelets and bladelet stalks are smooth (maximum bladelet count being 50-60 and the longest bladelet about 1 in.), it is a RATTLE BEAN.

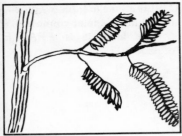

Fig. 94

Fig. 95

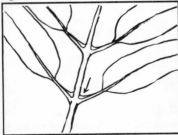

Fig. 96

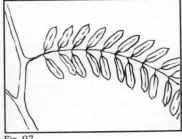

Fig. 97

PRAIRIE ACACIA

(Acacia hirta)

Ferny, low shrubs up to 3 ft. high with grooved stems. The tiny flowers are borne in rounded heads ½ in. in diameter. The fruit is a narrow pod 2-3 in. long. It is frequent in the grasslands of our five-state area. It belongs to the pea family *(Leguminosae)*.

LEADEN INDIGO-BUSH

(Amorpha canescens)

A low shrub up to 3 ft. tall which is densely gray-hairy throughout. See flower and fruit characteristics under INDIGO-BUSH, page 100. It grows in sandy woods and on stream banks throughout our area. It is a member of the pea family *(Leguminosae)*.

KIDNEYWOOD

(Eysenhardtia polystachya)

A shrub or small tree up to 20 ft. tall with many stems from its base. The pea flowers are small, white, hairy and glandular-dotted. The fruit is a pod ½ in. long and greenish-brown. It grows on dry slopes and ridges in New Mexico. It belongs to the pea family *(Leguminosae)*.

RATTLE BEAN

(Sesbania drummondii or *punicea)*

Shrubs up to 10 ft. tall with handsome pea flowers about 1 in. long. *S. drummondii* flowers have a yellow banner, whereas those of *S. punicea* have an orangish red or rose-colored banner. The fruit is an inflated, angled pod 2-3 in. long in which the loose seeds rattle when the wind blows. *S. drummondii* grows in low, wet ground in Texas, Arkansas and Louisiana. *S. punicea* is native to Brazil and escapes cultivation in Louisiana and eastern Texas. They belong to the pea family *(Leguminosae)*.

104

1. If the leaves form a radiating mass and are spine-tipped, go to (KI 54), in the left-hand margin of page 107.

2. If all of the leaves are opposite or whorled on the stem, as diagrammed to the right, find this drawing (KI 56), in the left-hand margin of page 111.

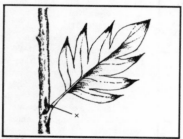

KI 53

3. If all of the leaves are alternate on the stem as pictured to the right, look for this drawing (KI 90), in the left margin of page 179.

4. If some of the leaves are opposite or whorled and some are alternate on your specimen, find the drawing to the right (KI 71), in the left-hand margin of page 141.

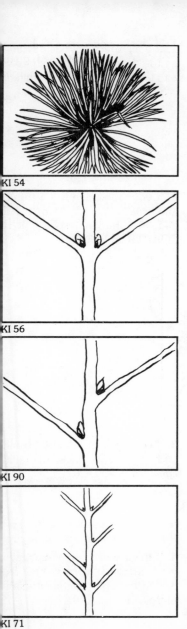

KI 54

KI 56

KI 90

KI 71

1. If your specimen has a trunk, and its leaves are 2-3 in. broad and lack spines on the margin, it is SPANISH DAGGER.

2. If your specimen has a trunk, and its leaves are ⅛-½ in. broad and lack marginal spines, it is SOAPTREE YUCCA.

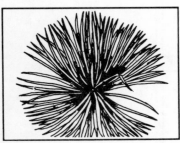

KI 54

3. If your specimen lacks a trunk and its leaves lack marginal spines, it is a YUCCA.

4. If the leaves of your specimen have marginal spines, find (KI 55), at the right, in the left-hand margin of page 109.

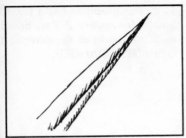

Fig. 98

Fig. 99

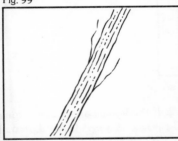

Fig. 100

KI 55

SPANISH DAGGER
(Yucca faxoniana, carnerosana or *treculeana)*
A group of three arborescent species to which this name is applied. Trees up to 18 ft. high with one or several trunks and branches terminated by leaf rosettes. The yellowish-white flowers are 2-3 in. long and the fruit is a 3-celled capsule 3-4 in. long. They grow in the desert areas of southwestern Texas and southern New Mexico, and belong to the lily family *(Liliaceae)*.

SOAPTREE YUCCA
(Yucca elata)
Trees up to 15 ft. tall with short branches terminated by leaf rosettes. Flowers white and 1-2 in. long with a dry capsule 1-2 in. long. It grows on desert plains and hills in Trans-Pecos Texas and the lower half of New Mexico. It belongs to the lily family *(Liliaceae)*.

YUCCA
(Yucca spp.)
A difficult-to-separate complex with a short trunk or only a woody root crown terminated by a leaf-rosette. The lily flowers (6 tepals and 6 stamens) are white to greenish, in showy terminal clusters, the fruit being a 3-celled, oblong capsule. They grow in mostly dry habitats throughout our 5-state area. They belong to the lily family *(Liliaceae)*.

1. If the leaves are 1 in. or less wide, it is SAWTOOTH SOTOL.

KI 55

2. If the leaves are 4-9 in. wide, it is an AGAVE or CENTURY-PLANT.

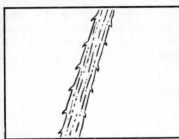

Fig. 101

SAWTOOTH SOTOL
(Dasylirion spp.)
A group of similar species with short trunks, spooned leaf bases, the leaves in a radial mass or rosette. The flowers are imperfect, tiny and in a central spike-like cluster that may be 2-15 ft. tall. The fruit is a tiny, winged capsule. They grow on dry soils in the southwestern part of our area. They belong to the lily family *(Liliaceae)*.

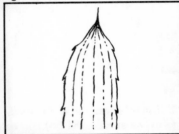

Fig. 102

CENTURY-PLANT
(Agave spp.)
Short-stemmed plants with large rosettes of fleshy leaves. Large flowers clustered at the top of a central, prominent scape with the fruit a 3-celled capsule. They grow on open, dry slopes in the southwestern part of Texas and southern New Mexico in our area. They belong to the amaryllis family
(Amaryllidaceae).

110

1. If your specimen possesses naked thorns or the twigs are thorn-tipped, find the picture (KI 57), shown at the right, in the left-hand margin of page 113.

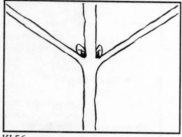

KI 56

2. If your specimen is thornless, go to the symbol (KI 58), shown to the right, in the left-hand margin of page 115.

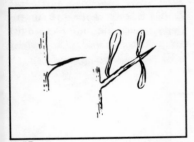

KI 57

KI 58

1. If the spines are naked and paired along the stem, it is CRUCILLA.

2. If the branchlets are thorn-tipped and the narrow leaves are no more than 3/5 in. long, it is BLACK-BUSH.

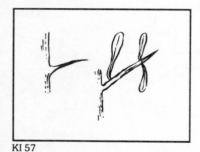

KI 57

3. If the branchlets are thorn-tipped and the leaves narrow-oblong, silvery-hairy with brown spots on their undersurface, it is SILVER BUFFALO-BERRY.

4. If the spines are naked on the twigs and the leaves are reverse tear-shaped and their undersurfaces wooly, it is WOOLY BUCKTHORN.

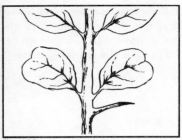

Fig. 103

CRUCILLA
(Randia rhagocarpa)
A shrub up to 6 ft. high with pale green flowers that are funnel-shaped with hairy throats and about 1/3 in. long. Fruit is rough-coated, globose and 1/2 in. in diameter. It grows in open brushlands in the lower Rio Grande valley of Texas. It belongs to the madder family *(Rubiaceae)*.

Fig. 104

BLACK-BUSH
(Coleogyne ramosissima)
A shrub up to 6 ft. tall with rigid diffuse branches. Flowers are greenish or purplish about 1/3 in. long with many stamens. The fruit is a small, flattened achene. It grows in sandy soils in New Mexico. It belongs to the rose family *(Rosaceae)*.

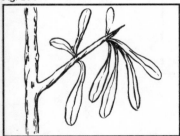

Fig. 105

SILVER BUFFALO-BERRY
(Shepherdia argentea)
A thicket-forming shrub to 15 ft. tall. Flowers imperfect and inconspicuous with oval, fleshy, edible, red or yellow fruits, 1/4 in. long. It grows in sandy plains and in canyons in New Mexico. It belongs to the oleaster family *(Eleagnaceae)*.

Fig. 106

WOOLY BUCKTHORN
(Bumelia lanuginosa)
A shrub or tree up to 45 ft. tall. The white flowers are 1/5 in. long and the fruit is roundish, lustrous black, fleshy and up to 1 in. long. It grows in upland or streamside forests in our area. It belongs to the sapodilla family *(Sapotaceae)*.

114

1. If the leaves of your specimen have smooth, entire margins, go to the drawing (KI 76), shown at the right, in the left-hand margin of page 151.

2. If the leaves of your specimen possess 3-5 lobes, as shown at the right (KI 59), go to the left-hand margin of page 117.

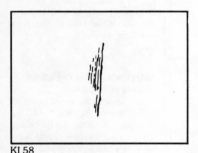

KI 58

3. If the leaves of your specimen are not lobed and have sharp-toothed or blunt-toothed margins, as shown to the right (KI 63), go to this symbol in the left-hand margin of page 125.

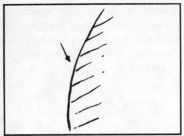

KI 76

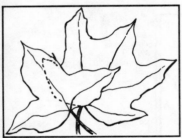

KI 59

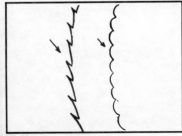

KI 63

1. If the lobes are toothed, as shown to the right (KI 61), go to this picture in the left-hand margin of page 121.

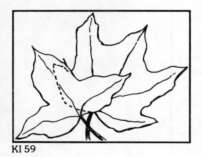

KI 59

2. If the lobes are smooth, as shown to the right, go to this drawing (KI 60), in the left-hand margin of page 119.

117

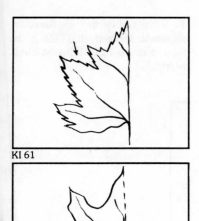

KI 61

KI 60

1. If the leaf is sparsely hairy on its undersurface and 1½-3 in. long, it is the FLORIDA SUGAR MAPLE.

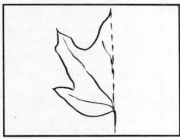

KI 60

2. If the leaf is soft-hairy on its underneath surface and 5-6 in. long, it is the BLACK MAPLE.

Fig. 107

Fig. 108

FLORIDA SUGAR MAPLE
(Acer barbatum)
A tree up to 60 ft. tall, its young bark being smooth and whitish. Its flowers are inconspicuous and its fruit consists of two paired samaras with green to reddish wings ¾ in. long. It grows in moist, rich soil in Louisiana, Arkansas, eastern Oklahoma and southeastern Texas. It belongs to the maple family *(Aceraceae)*.

BLACK MAPLE
(Acer nigrum)
A tree up to 120 ft. tall with a spreading crown and pale gray, smooth, young bark becoming dark and deeply fissured. The flower is inconspicuous and the fruit consists of paired samaras with reddish-brown wings ½-1½ in. long. It grows in southwestern Arkansas and Louisiana. It belongs to the maple family *(Aceraceae)*.

120

1. If the leaf is somewhat hairy beneath and the lobe teeth blunted, it is BIG-TOOTH MAPLE.

2. If the leaf is hairy beneath and the lobe teeth are sharp, it is MAPLE-LEAF VIBURNUM.

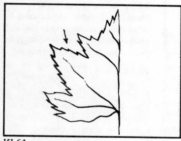

KI 61

3. If the leaf is non-hairy, go to the symbol, shown to the right (KI 62), in the left-hand margin of page 123.

121

Fig. 109

BIG-TOOTH MAPLE
(Acer grandidentatum)

A tree up to 45 ft. tall with an open, rounded crown and dark brown to gray bark with narrow fissures. The flower is inconspicuous and the fruit consists of paired samaras with green or rose-colored wings ¾-1 in. long. It grows in valleys along streams in the Edward's Plateau and Trans-Pecos mountains of Texas, Wichita mountains of Oklahoma and the Organ, White and Sacramento mountains of New Mexico. It belongs to the maple family *(Aceraceae)*.

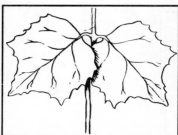

Fig. 110

MAPLE-LEAF VIBURNUM
(Viburnum acerifolium)

A slender-branched shrub up to 6 ft. tall. Its small flowers are ¼ in. wide, cup-shaped and creamy-white. The fruit is fleshy, ovoid, purplish-black and ⅓ in. long. It grows in sandy soils in eastern Texas, Arkansas and Louisiana. It belongs to the honeysuckle family *(Caprifoliaceae)*.

non-hairy

KI 62

122

1. If the leaf length is 1-3 in. it is the ROCKY MOUNTAIN MAPLE.

2. If the leaf length is 6-7 in. and the lobe teeth few and blunt, as shown to the right, it is SUGAR MAPLE.

non-hairy

KI 62

3. If the leaf length is 3-6 in., the lobe teeth sharp and the lobes deep-cut, as shown to the right, it is SILVER MAPLE.

4. If the leaf length is 3-6 in., the lobe teeth sharp and the lobes shallow-cut, as shown to the right, it is RED MAPLE.

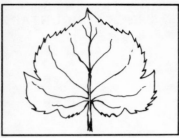
Fig. 111

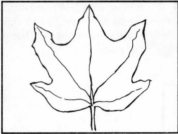
Fig. 112

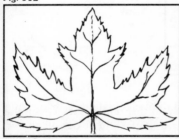
Fig. 113

Fig. 114

ROCKY MOUNTAIN MAPLE
(Acer glabrum)
A small tree up to 30 ft. tall with smooth, gray bark and an irregular crown. The flowers are inconspicuous and the fruit consists of paired samaras with tan wings up to 1½ in. long. It grows in open seepage areas in the mountains of New Mexico. It belongs to the maple family *(Aceraceae)*.

SUGAR MAPLE
(Acer saccharum)
A large tree up to 120 ft. tall with smooth, gray bark when young, becoming fissured with curly plates. The flowers are inconspicuous and the fruit consists of paired samaras with reddish-brown wings 1 in. long. It grows on forested hills and ravines in Arkansas, eastern Oklahoma and northeastern Texas. It belongs to the maple family *(Aceraceae)*.

SILVER MAPLE
(Acer saccharinum)
A tree up to 100 ft. tall with a rounded crown. The flowers are inconspicuous and the fruit is a pair of samaras with 1 in. long wings. It grows in rich forests in Arkansas, Louisiana and eastern Oklahoma and Texas. It belongs to the maple family *(Aceraceae)*.

RED MAPLE
(Acer rubrum)
A large-crowned tree up to 100 ft. tall with smooth, red branchlets. The flowers are inconspicuous and the fruit consists of paired samaras with yellowish-green to brown wings ½-1 in. long. It grows in moist ground along streams in Arkansas and Louisiana and eastern Texas and Oklahoma. It belongs to the maple family *(Aceraceae)*.

1. Turn over a leaf, if non-hairy, as shown to the right, go to this symbol (KI 64), in the left-hand margin of page 127.

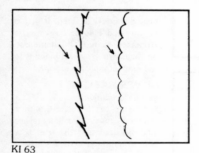

KI 63

2. If the under surface of the leaf is sparsely or densely hairy, as shown to the right, go to this symbol (KI 66), in the left-hand margin of page 131.

non-hairy

KI 64

hairy

KI 66

1. If the leaves are tear-shape and appear stalkless, it is STRAWBERRY-BUSH.

2. If the leaves are tear-shape and possess stalks, it is DESERT OLIVE.

KI 64

3. If the leaves are oval to elliptic, as shown to the right (KI 65), proceed to this drawing in the left-hand margin of page 129.

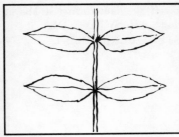

Fig. 115

STRAWBERRY-BUSH
(Euonymus americanus)
Erect shrub up to 6 ft. tall, sometimes straggling. The small flowers are greenish-purple and the fruit is a crimson capsule when ripe. It grows in rich, moist forests of Arkansas, Louisiana, eastern Texas and Oklahoma in our area. It belongs to the bittersweet family *(Celastraceae)*.

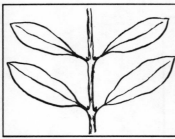

Fig. 116

DESERT OLIVE
(Forestiera pubescens var. *glabrifolia)*
Erect shrubs or small trees up to 12 ft. tall. The small flowers lack petals and are inconspicuous. The fruit is a small, bluish-black olive. It grows in brushy prairies in Trans-Pecos Texas and New Mexico in our area. It belongs to the olive family *(Oleaceae)*.

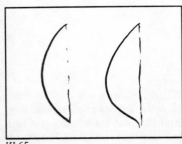

KI 65

1. If the leaves are ¼-1¼ in. long, it is PACHYSTIMA.

2. If the leaves are 1½-4 in. long, and the leaf stalk is red-hairy, it is RUSTY VIBURNUM.

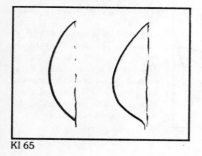

KI 65

3. If the leaves are 1½-4 in. long and the leaf stalk lacks the reddish hairs, it is BLACKHAW.

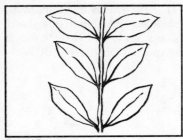
Fig. 117

PACHYSTIMA
(Pachystima myrsinites)
A dense, evergreen shrub up to 3 ft. tall. Its flowers are minute and its fruit is a tiny, leathery capsule. It grows in the western mountains of Texas and New Mexico from about 6,000 ft. up to timberline. It belongs to the bittersweet family *(Celastraceae)*.

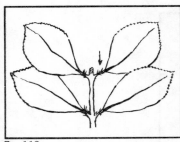
Fig. 118

RUSTY VIBURNUM
(Viburnum rufidulum)
A tall shrub or small tree up to 30 ft. tall with blackish, checkered bark. The white flowers are small, trumpet-shaped, with protuding stamens and arranged in flat clusters. The fruits are flattish, prominent, bluish-black olives (drupes). It is common at the edge of woods or in streamside thickets in all states of our area except New Mexico. It belongs to the honeysuckle family *(Caprifoliaceae)*.

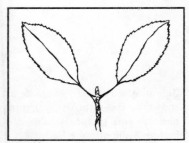
Fig. 119

BLACKHAW
(Viburnum prunifolium)
Identical to RUSTY VIBURNUM above except the leaf stalks lack the red hairs. It also has the same range and habitat as the RUSTY VIBURNUM.

130

1. If the leaves are apparently stalkless and marginal teeth sharp and oval to elliptic in shape, it is DEVIL'S ELBOW.

2. If the leaves are apparently stalkless, their shape narrowly oblong and the marginal teeth blunt, it is BUTTERFLY-BUSH.

KI 66

3. If the leaves are stalked, as shown to the right (KI 67), proceed to this picture in the left-hand margin of page 133.

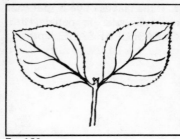
Fig. 120

DEVIL'S ELBOW
(Forestiera pubescens)
Mostly a straggly shrub or sometimes a small tree up to 15 ft. tall. The small flowers lack petals and are inconspicuous. The fruit is a small bluish-black olive. It grows in wet, rich soil along streams in New Mexico, Texas, and Oklahoma. It belongs to the olive family *(Oleaceae)*.

Fig. 121

BUTTERFLY-BUSH
(Buddleja scordioides)
A densely-branched, aromatic shrub with all parts rusty-hairy, growing to 4 ft. tall. The minute white flowers are clustered into small, hairy heads and the fruit is inconspicuous. It grows in open sunny sites in New Mexico and southwestern Texas. It belongs to the logania family *(Loganiaceae)*.

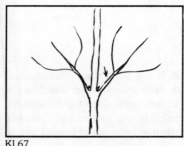
KI 67

132

1. If the leaves of your specimen are triangular in outline, it is GREASELEAF SAGE.

2. If the leaves of your specimen are elliptic, oval or tear-shaped in outline and some of them are more than 5-9 in. in length, it is FRENCH-MULBERRY.

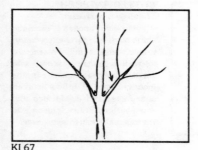

KI 67

3. If the leaves are elliptic, oval or tear-shaped and less than 5 in. in length, go to the symbol (KI 68), shown to the right, in the left-hand margin of page 135.

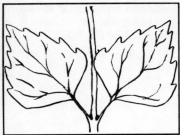

Fig. 122

GREASELEAF SAGE
(Salvia pinquifolia)
An aromatic, white-hairy shrub up to 5 ft. tall. The blue tubular flowers are borne in terminal spikes and the fruit is a group of four nutlets clustered in the persistent calyx base. It is found on dry mesas and hillsides in New Mexico and westernmost Texas. It belongs to the mint family *(Labiatae)*.

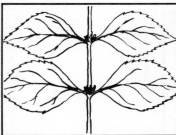

Fig. 123

FRENCH-MULBERRY
(Callicarpa americana)
A shrub with hairy branches up to 9 ft. tall. Its tiny tubular flowers are blue to white and the prominent purplish fruits are clustered around the leaf stalks. It grows in woods and thickets on moist, rich soil in all states of our area except New Mexico. It belongs to the verbena family *(Verbenaceae)*.

⟨ 5″

KI 68

134

1. Measure several leaves of your specimen, if all are ⅛-⅝ in. long and the stems smooth, it is ALOYSIA.

KI 68

2. If some of the leaves are over an inch in length, go to this symbol (KI 69), in the left-hand margin of page 137.

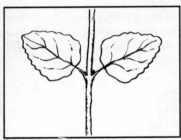

Fig. 124

1" ⟩

KI 69

ALOYSIA

(Aloysia wrightii)

A sweet-aromatic shrub with hairy branches up to 6 ft. tall. The many, minute, densely-clustered flowers are trumpet-shaped, hairy and white. The fruit consists of two tiny nutlets. It grows in rocky canyons or rocky slopes in New Mexico and southwestern Texas. It belongs to the verbena family *(Verbenaceae)*.

136

1. If the young twig is square or 4-angled, go to the picture (KI 70), shown to the right, in the left-hand margin of page 139.

2. If the young twig is round and thus not angled and the marginal teeth blunted, it is CLIFF JAMESIA.

KI 69

3. If the twig is round, the marginal teeth sharp and the leaf stalk hairy, it is HAIRY MOCKORANGE.

4. If the twig is round, the marginal teeth sharp and the leaf stalk mostly smooth, it is the SOUTHERN ARROWWOOD.

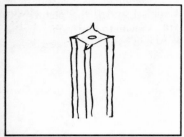

KI 70

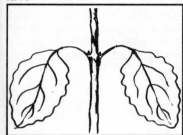

Fig. 125

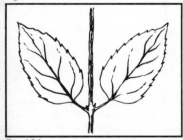

Fig. 126

Fig. 127

CLIFF JAMESIA
(Jamesia americana)
A shreddy-barked shrub up to 6 ft. tall. Its white-hairy flowers are borne in flat clusters and the fruit is inconspicuous. It is found in the high mountains of New Mexico in our area on dry, well-drained soil. It belongs to the saxifrage family *(Saxifragaceae)*.

HAIRY MOCKORANGE
(Philadelphus hirsutus)
A straggly, brown shreddy-barked shrub up to 6 ft. tall with large white flowers possessing many stamens. The fruit is a small capsule. It occurs along streamsides or rocky slopes in Arkansas in our area. It belongs to the saxifrage family *(Saxifragaceae)*.

SOUTHERN ARROWWOOD
(Viburnum dentatum)
A usually multistemmed shrub up to 15 ft. tall with slender, hairy branches. The small, white, tubular flowers are arranged in flat clusters and the fruit is a bluish-black olive (drupe) about 2/5 in. long. It grows mostly in moist, sandy soil in Louisiana, Arkansas and eastern Texas in our area. It belongs to the honeysuckle family *(Caprifoliaceae)*.

1. If some of the longer leaves are 3-5 in., it is EASTERN WAHOO.

2. If the longest leaves are under 3 in. and the stem is rough-prickly, it is LANTANA.

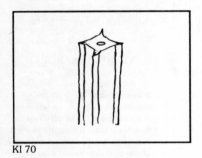

KI 70

3. If the leaves are 1 to 2¾ in. long, the stems smooth and the leaf stalk very hairy, it is RED-BUSH.

4. If the leaves are ⅓-1½ in. long, the stems smooth and the leaf stalk smooth, it is BLUE SAGE.

Fig. 128

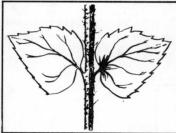

Fig. 129

Fig. 130

Fig. 131

EASTERN WAHOO
(Euonymus atropurpureus)
A shrub or small tree up to 25 ft. tall with purplish-green twigs. The flowers are small, purplish becoming 3-4 lobed capsules that are purple or red and contain red seeds. In our area it grows in rich woodlands or ravine thickets in Arkansas and eastern Texas. It belongs to the bittersweet family *(Celastraceae)*.

LANTANA
(Lantana spp.)
A complex of similar species as irregularly-shaped, often aromatic, low shrubs up to 6 ft. tall. Small flowers occur in flat clusters with corolla varigation in the same cluster from yellow through orange to red. The fruit is drupe-like, bearing two nutlets. Most species occur on sandy soil throughout our area. They belong to the verbena family *(Verbenaceae)*.

RED-BUSH
(Lippia graveolens)
A slender, aromatic shrub up to 9 ft. tall or a small tree up to 27 ft. tall. The small, tubular, yellow-white flowers are borne in the leaf-stalk angles. The fruit is a dry olive (drupe). It grows on dry, rocky hills in New Mexico and western Texas. It belongs to the verbena family *(Verbenaceae)*.

BLUE SAGE
(Salvia ballotaeflora)
An aromatic, many-branched shrub up to 6 ft. tall. The handsome blue flowers are tubular and two-lipped. The fruit consists of 4 nutlets within the folded calyx. It grows on dry limestone hills of central Texas westward into New Mexico. It belongs to the mint family *(Labiatae)*.

140

1. If the leaves of your specimen have apparent leaf stalks, go to the picture, shown to the right (KI 73), in the left-hand margin of page 145.

2. If the leaf stalks are apparently lacking and your specimen is thorny, it is WOLFBERRY.

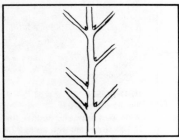

KI 71

3. If your specimen lacks leaf stalks and is not thorny, go to the picture (KI 72), shown at the right, in the left-hand margin of page 143.

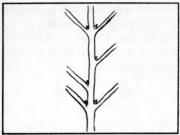

KI 73

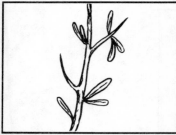

Fig. 132

WOLFBERRY
(Lycium spp.)
These thorny, spatulate-leaved shrubs are similar in appearance, mostly scraggly and ranging from 1 to 9 ft. tall. Their flowers are long-tubular and range in color from white to purple to blue. The fruit is an orange to red berry. Two species grow in western Texas and New Mexico in dry habitats and a third species is found in saline flats in southern Texas and Louisiana along the coast. They belong to the nightshade family *(Solanaceae)*.

KI 72

1. If the leaves are silvery-hairy, it is a SAGE-BRUSH.

2. If the leaves are smooth and narrow and 4-12 in. long, it is DESERT-WILLOW.

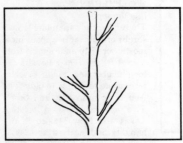

KI 72

3. If the leaves are hairy, but not silvery, narrow and ½-1¾ in. long, it is WINTER FAT.

Fig. 133

SAGE-BRUSH
(Artemesia spp.)
A shrubby complex of 6 similar species in our area. Most are low-growing, although the BIG SAGE-BRUSH may reach 10 ft. tall. The aster type of flowers are small and borne in a panicle type of cluster rather than the head as in most composites. The small seed-like fruit (achene) is 4-5 ribbed. The sagebrushes grow in dry prairies of the Great Plains of New Mexico and western Texas and Oklahoma in our area. They belong to the aster family *(Compositae)*.

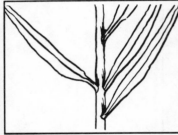
Fig. 134

DESERT-WILLOW
(Chilopsis linearis)
A lax shrub or tree with slender branches, growing up to 30 ft. tall. The flowers are showy, tubular, 2-lipped, pink with purple-streaked throats and 1-1½ in. long. The fruit is a slender pod 4 to 12 in. long. The seeds within the pod are distinctively flattened and winged. It grows in desert washes in Texas and New Mexico in our area. It belongs to the bignonia family *(Bignoniaceae)*.

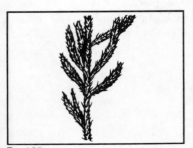
Fig. 135

WINTER FAT
(Eurotia lanata)
Wooly shrub that is many-branched from the base, growing up to 3 ft. tall. The flowers are indistinctively small and greenish as are the small, 2-horned, dry fruits. It is found on dry mesas and plains of Texas and New Mexico in our area. It belongs to the goosefoot family *(Chenopodiaceae)*.

144

1. If thorniness is lacking in your specimen, it is GOLDENEYE.

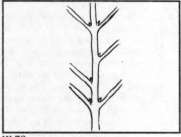

KI 73

2. If naked thorns are present or the twigs are thorn-tipped, go to the picture shown to the right (KI 74), in the left-hand margin of page 147.

Fig. 136

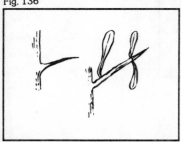

KI 74

GOLDENEYE
(Viguiera deltoidea var. *parishii)*
A many-branched shrub with slender, grayish branches, growing up to 3 ft. tall. Flowers grouped into heads and resembling sunflowers. Ray or outer flowers yellow. Fruit small, seedlike. It grows in the deserts of New Mexico. It belongs to the aster family *(Compositae)*.

1. If the leaves are distinctively lobed and saw-toothed as shown to the right, it is a CURRANT.

2. If the leaves are unlobed, narrow and 1/12-¼ in. long, it is JAVELINA BUSH.

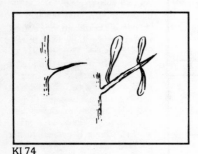

KI 74

3. If the leaves are unlobed and longer than ¼ in. find the picture (KI 75), shown to the right, in the left-hand margin of page 149.

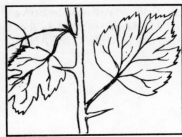

Fig. 137

CURRANT
(Ribes spp.)
A group of similar and difficult-to-separate species of mostly low shrubs with small flowers and juicy berries. They occur throughout our area in mostly forested habitats. They belong to the saxifrage family *(Saxifragaceae)*.

Fig. 138

JAVELINA BUSH
(Condalia ericoides)
A densely-branched, sprawling, evergreen shrub up to 3 ft. tall. The yellowish flowers are inconspicuous and the fruit is a black olive (drupe) ⅓ in. long. It occurs mostly in the desert habitats of southwestern Texas and New Mexico, extending onto the plains of Texas to the Edwards Plateau. It belongs to the buckthorn family *(Rhamnaceae)*.

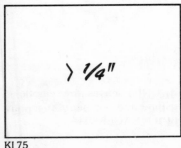

> ¼″

KI 75

148

1. If the leaves are narrow and non-hairy, it is SNAKE-EYES.

2. If the leaves are narrow and silvery-hairy below, it is GOAT-BUSH.

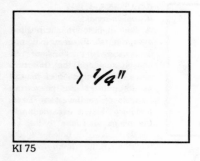

> 1/4"

KI 75

3. If the leaves are elliptic in outline and non-hairy at maturity, it is SAGERETIA.

Fig. 139

SNAKE-EYES
(Phaulothamnus spinescens)
A diffusely branched shrub up to 8 ft. tall with gray, divaricate branches. Flowers tiny, the fruit small, fleshy, sphaerical and whitish. Uncommon in the Rio Grande Valley of Texas on clay soils. It belongs to the pokeweed family *(Phytolaccaceae)*.

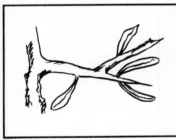

Fig. 140

GOAT-BUSH
(Castela texana)
A low, densely-branched shrub up to 6 ft. tall. Flowers small, red or orange; fruit a brilliant red olive (drupe) ⅓ in. long. It occurs on gravelly soils in thickets and on mesquite prairies of central, southwestern and western Texas. It belongs to the quassia family *(Simaroubaceae)*.

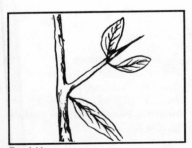

Fig. 141

SAGERETIA
(Sageretia wrightii)
A slender, straggling shrub, mostly 2-5 ft. tall with rigid, divaricate branches. The flower is small and inconspicuous, but the fruit is a black olive (drupe). It occurs on dry, rocky, mountain slopes in the mountains of Trans-Pecos Texas. It belongs to the buckthorn family *(Rhamnaceae)*.

150

1. If your specimen grows from the wood of branches of other trees and shrubs as a parasite, it is a MISTLETOE.

2. If your specimen grows from the ground and its leaves are apparently stalkless, proceed to (KI 77), shown to the right, in the left-hand margin of page 153.

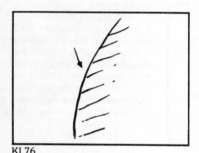

KI 76

3. If your specimen grows from the ground and its leaves are apparently stalked, look for the picture (KI 82), shown to the right, in the left-hand margin of page 163.

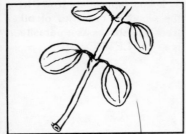

Fig. 142

MISTLETOE
(Phoradendron spp.)
A group of semi-woody, green-branched, parasitic shrubs up to 2 ft. tall. The flower is inconspicuous and small; the fruit is a whitish, sticky berry. It occurs throughout our area. It belongs to the mistletoe family *(Viscaceae)*.

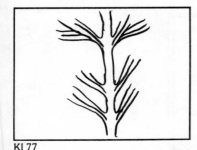

KI 77

KI 82

152

1. If your specimen is low-spreading on sandy beaches with rounded, fleshy leaves, it is BEACHWORT.

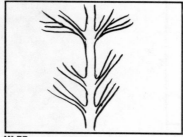

KI 77

2. If your specimen has thin, non-fleshy leaves proceed to (KI 78), shown to the right, in the left-hand margin of page 155.

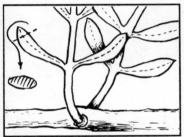

Fig. 143

BEACHWORT
(Batis maritima)
A creeping, shrub-like plant with erect shoots and strong-scented, fleshy, waxy leaves. Flowers imperfect, small and inconspicuous. The waxy-green fruit is fleshy and up to ¾ in. long. It is found on sandy beaches and mud flats of the seashores of Texas and Louisiana of our area. It belongs to the saltwort family *(Bataceae)*.

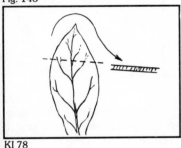

KI 78

154

1. If the leaves are non-hairy and smooth, proceed to the picture (KI 79), shown to the right, in the left-hand margin of page 157.

2. If the leaves are reverse tear-shape and densely hairy, it is MEXICAN FIDDLEWOOD.

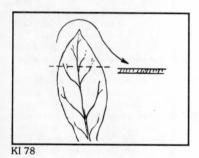

KI 78

3. If the leaves are oval to elliptic in outline and white-hairy, it is CENIZO.

4. If the leaves are sparsely-hairy, rough to the touch and oblong, as shown to the right, it is SCARLET BOUVARDIA.

155

KI 79

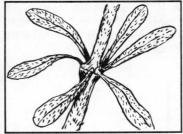

Fig. 144

Fig. 145

Fig. 146

MEXICAN FIDDLEWOOD
(Citharexylum brachyanthum)

A stiff, irregularly branched shrub up to 8 ft. tall with corky ridges in the branch angles. The irregular, tubular flower is small and white; the fruit is rounded, fleshy and shiny red at maturity. It grows on dry, rocky hills and in valleys of western Texas. It belongs to the verbena family *(Verbenaceae)*.

CENIZO
(Leucophyllum frutescens)

A shrub up to 8 ft. tall that is densely white-hairy on all parts. Its tubular flowers are showy, hairy, pale violet to pink or rarely white and up to 1 in. long. The fruit is a dry capsule about 1/5 in. long. It grows on rocky limestone hills in central, western and southwestern Texas and New Mexico. It belongs to the figwort family *(Scrophulariaceae)*.

SCARLET BOUVARDIA
(Bouvardia ternifolia)

A shrub up to 3 ft. tall with gray, divaricate branches. The attractive, slender-tubed flowers are red and mostly 1½ in. long and borne in terminal clusters. The fruit is a small capsule bearing winged seeds. It occurs in dry mountains among rocks in New Mexico and western Texas. It belongs to the madder family *(Rubiaceae)*.

156

1. If the longest leaves are less than ½ in. long and the twig is round, it is FRANKENIA.

2. If the longest leaves are less than ½ in. long and the twig is squarish (angled), it is SHRUB BLUET.

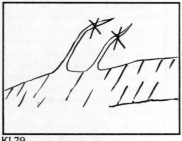

KI 79

3. If the longest leaves are 1 in. or longer, proceed to drawing (KI 80), at the right, in the left-hand margin of pages 159 and 59A.

157

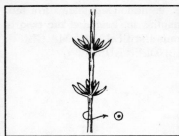

Fig. 147

FRANKENIA
(Frankenia jamesii)
A small, highly branched shrub up to 2 ft. tall with jointed stems. The flower is small and white. The fruit is a small, inconspicuous capsule. It grows on soils rich in gypsum, as alkaline salt flats in New Mexico and the plains country and Trans-Pecos of Texas. It belongs to the frankenia family *(Frankeniaceae)*.

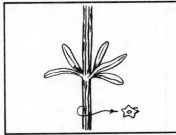

Fig. 148

SHRUB BLUET
(Hedyotis intricata)
A diffuse shrub up to 2 ft. high with stiff, grayish branches. The tiny flowers are white in small, flat clusters. The fruit is a top-shaped capsule 1/16 in. long. It grows on dry, open areas in New Mexico and Trans-Pecos Texas. It belongs to the madder family *(Rubiaceae)*.

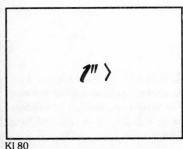

KI 80

158

1. If the twigs are grooved lengthwise, and the leaves are long-elliptical, it is CLIFF FENDLER-BUSH.

2. If the twigs are not grooved and all of the leaves are tear-shaped, it is DESERT-HONEYSUCKLE.

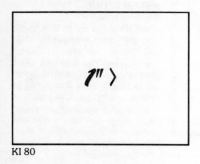

KI 80

3. If the twigs are grooved and the leaves are narrow, as shown to the right, it is FLAXLEAF BOUCHEA.

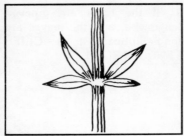

Fig. 149

CLIFF FENDLER-BUSH
(Fendlera rupicola)
Highly-branched shrub up to 6 ft. tall with slender, arching branches. The white flowers are small and the fruit is a capsule crowned with 3 beaks. It grows on rocky ledges and slopes of the mountains in New Mexico and western Texas. It belongs to the saxifrage family *(Saxifragaceae)*.

Fig. 150

DESERT HONEYSUCKLE
(Anisacanthus thurberi)
A shrub up to 8 ft. tall with shreddy, white bark peeling off in thin strips. The flower is showy, funnel-shaped, 2-lipped and about 1½ in. long, ranging in color from yellow to orange-red or sometimes bluish. The fruit is a dry, shiny capsule. It grows on dry mountain slopes, mesas and hills of southwestern New Mexico. It belongs to the acanthus family *(Acanthaceae)*.

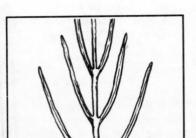

Fig. 151

FLAXLEAF BOUCHEA
(Bouchea linifolia)
A many-branched low shrub up to 3 ft. tall. The flowers occur in elongate, terminal clusters, each being trumpet-shaped, irregularly lobed, white to purple in color and about ¾ in. long. The fruit consists of a pair of beaked, small nuts. It occurs on dry, rocky hillsides in western Texas. It belongs to the verbena family *(Verbenaceae)*.

4. If the twigs are smooth and some leaves narrow and some reverse tear-shaped, it is FLAX-LEAF ST. PETER'S-WORT.

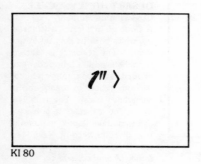

KI 80

5. If the twigs are smooth and the leaves oval to elliptic in outline, proceed to (KI 81), on the right, in the left-hand margin of page 161.

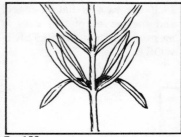

Fig. 152

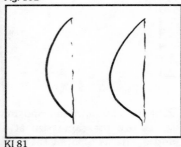

KI 81

FLAXLEAF ST. PETER'S-WORT

(Ascyrum hypericoides)

A leafy, low shrub with shreddy-barked twigs, growing up to 2 ft. tall. Flowers prominent, yellow and spidery-looking due to their many stamens. The fruit is a dry, beaked capsule about ⅓ in. long. It grows in sandy soil, in the eastern third of Texas, Oklahoma, Arkansas and Louisiana. It belongs to the St. John's-wort family *(Hypericaceae)*.

1. If the twig is flat (2-edged) and the leaves of two sizes in most whorls, it is ST. PETER'S-WORT.

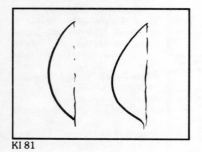

KI 81

2. If the twig is round and the leaves are the same size at a node, it is SMALL-LEAF MOCKORANGE.

161

Fig. 153

ST. PETER'S-WORT
(Ascyrum stans)
A shreddy-barked shrub up to 3 ft. tall. Flowers large, showy, bright yellow, stamens numerous; fruit a dry capsule. It grows in sandy bogs, meadows and forests in Oklahoma, Arkansas, Louisiana and eastern Texas. It belongs to the St. Peter's-wort family *(Hypericaceae)*.

Fig. 154

SMALL-LEAF MOCKORANGE
(Philadelphus microphyllus)
A delicate-branched shrub up to 6 ft. tall with fragrant, white flowers 1 in. across. The four petals have tips that appear gnawed. The fruit is a dry capsule. It occurs on dry, exposed, rocky slopes in the mountains of New Mexico and western Texas. It belongs to the saxifrage family *(Saxifragaceae)*.

1. If your specimen is growing in a tidal flat, it is BLACK-MANGROVE.

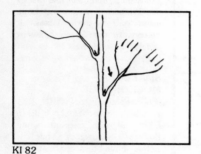

KI 82

2. If your specimen is rooted in soil, or in a fresh-water swamp (not in salt water), find the picture to the right, (KI 83), in the left-hand margin of page 165.

163

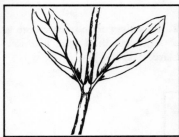

Fig. 155

BLACK-MANGROVE

(Avicennia germinans)
A shrub up to 3 ft. tall in our
area. The flower is bell-shaped,
¾ in. long, white with a yellow
throat and hairy tube. The fruit is
a hairy, beaked capsule about
1½ in. long. It grows in shallow
estuaries along the coast of Louisiana and southeastern Texas in
our area. It belongs to its own
family, the black-mangrove family *(Avicenniaceae)*.

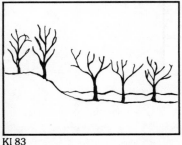

KI 83

1. Look at a young twig, if it is squarish and the leaves are densely hairy on their underneath surface, it is BEE-BRUSH.

2. If the twig of your specimen is squarish but the leaves are non-hairy, it is AUTUMN SAGE.

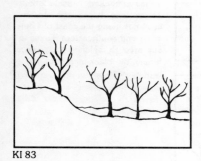

KI 83

3. If the twig of your specimen is rounded, proceed to the picture (KI 84), at the right, in the left-hand margin of pages 167 and 167A.

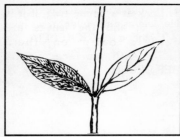

Fig. 156

BEE-BRUSH
(Aloysia gratissima)
An aromatic, thicket-forming, slender shrub up to 9 ft. tall. Small, 2-lipped, tubular flowers that are white to violet in color. The fruit is small and dry. It grows in dry habitats in New Mexico and western Texas. It belongs to the verbena family *(Verbenaceae).*

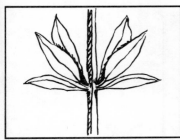

Fig. 157

AUTUMN SAGE
(Salvia greggii)
A slender-branched shrub up to 5 ft. tall. The red or red-purplish tubular flowers are 2-lipped and 1 in. long. The fruit is a nest of 4 nutlets. It grows in dry, exposed sites in New Mexico and western Texas. It belongs to the mint family *(Labiatae).*

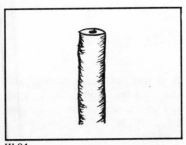

KI 84

166

1. If your specimen has thread-like leaves and never grows over 2 ft. high, it is BRICKELLIA.

2. If your specimen has hairy-scaly leaf stalks and grows only in acid swamps, it is POSSUM-HAW.

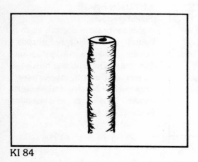

KI 84

3. If the petioles are hairy and the blades hairy and your specimen grows on limestone hills, it is MEXICAN SILK-TASSEL.

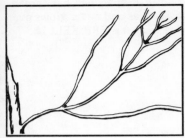

Fig. 158

BRICKELLIA
(Brickellia squamulosa)
A low shrub up to 2 ft. tall. The asterlike flowers are white and borne in ½ in. high heads. The fruit is a very small, seed-like achene. It grows on dry, exposed soil in the lower mountain reaches of southwestern New Mexico. It belongs to the aster family *(Compositae)*.

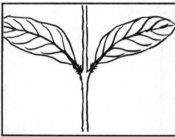

Fig. 159

POSSUM-HAW
(Viburnum nudum)
A shrub or small tree up to 15 ft. tall with leathery, shiny leaves. The flowers are white, broadly bell-shaped, ⅓ in. wide and borne in terminal, flattened clusters. The fruit is a bluish-black olive (drupe), ¼ in. in diameter. It grows in acid swamps of mostly pinelands in Arkansas, Oklahoma, Louisiana and eastern Texas. It belongs to the honeysuckle family *(Caprifoliaceae)*.

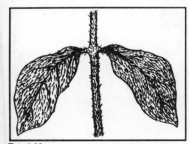

Fig. 160

MEXICAN SILKTASSEL
(Garrya ovata)
An evergreen shrub or small tree up to 12 ft. tall, covered with curled, brownish hairs. Male and female flowers inconspicuous, the bluish-purple drupes ⅓ in. long. It grows on limestone outcrops in central and western Texas. It belongs to the dogwood family *(Cornaceae)*.

168

4. If the petioles are hairy and the blades smooth and your specimen is in the mountains, it is the WRIGHT SILKTASSEL.

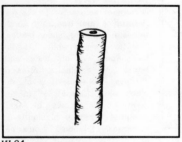

KI 84

5. If your specimen has broad leaves and smooth leaf stalks, go to the drawing (KI 85), at the right, in the left-hand margin of page 169.

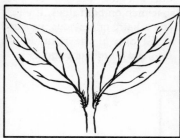

Fig. 161

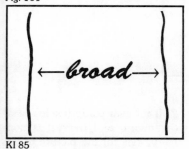

KI 85

WRIGHT SILKTASSEL

(Garrya wrightii)

An evergreen shrub or small tree up to 15 ft. tall. Flowers and fruit similar to those of the MEXICAN SILKTASSEL (see above). It grows on rocky outcrops in the mountains of New Mexico and extreme western Texas. It belongs to the dogwood family *(Cornaceae)*.

1. Examine the blade of a leaf (but not the nerves) of your specimen on both sides. If either side is hairy or resinous, proceed to (KI 86), at the right, in the left-hand margin of page 171.

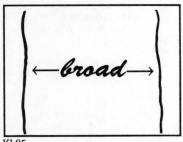

KI 85

2. If the leaf blade (exclusive of the midrib and nerves) is non-hairy, look for the picture to the right (KI 88), in the left-hand margin of page 175.

hairy

KI 86

non-hairy

KI 88

1. If the leaves are resinous (and also small up to 2/5 in. long), it is CREOSOTE BUSH.

2. If the leaves are large (6-12 in. long), and heart-shaped, it is the CATALPA.

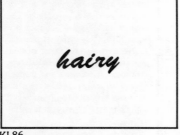

KI 86

3. If the leaf-stalks are grooved, it is the FLOWERING DOGWOOD.

4. If the leaves of your specimen are non-resinous nor very large (more than 8 in. long), nor have channeled leaf-stalks, proceed to the drawing (KI 87), pictured to the right, in the left-hand margin on page 173.

Fig. 162

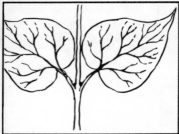

Fig. 163

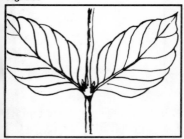

Fig. 164

CREOSOTE BUSH
(Larrea tridentata)
An evergreen shrub up to 9 ft. tall, with many flexuous stems from its base. The evergreen leaves are sticky with resin and smell of creosote, particularly when wet. The flower is yellow and small. The fruit is a small, white-silky capsule. It is probably the most common shrub of desert flats in New Mexico and western Texas. It belongs to the caltrop family *(Zygophyllaceae)*.

CATALPA
(Catalpa speciosa)
A tree, 50 to 90 ft. tall. The flower is strikingly bell-shaped, 2-lipped, the lips ruffled, white, the throat yellow and purple-blotched. Flower length is 2 in. and its width is about 2½ in. The fruit is an elongate, slender pod from 8 to 20 in. long. It grows in moist, rich woods, along streams and around ponds in Oklahoma, Arkansas, Louisiana and eastern Texas. Catalpa belongs to the bignonia family *(Bignoniaceae)*.

FLOWERING DOGWOOD
(Cornus florida)
A shrub or tree up to 40 ft. tall with a spreading crown. The "flower" is actually a cluster with 4 prominent white bracts, each 2 in. long. The flower is inconspicuous and greenish-white. The fruits are oval, red olives about ½ in. long. It grows in rich woods in eastern Oklahoma, Arkansas, Louisiana and eastern Texas. It belongs to the dogwood family *(Cornaceae)*.

1. If the leaves are ovate-elliptic, ½-5 in. long and aromatic when crushed, it is CAROLINA ALL-SPICE.

2. If the leaves are narrow ovate-elliptic and ¼-¾ in. long, it is THYME-LEAVED MOCK-ORANGE.

KI 87

3. If the leaves are ovate and 2-8 in. long, it is BUTTONBUSH.

4. If the leaves are ovate and ½-2½ in. long, it is CANADIAN BUFFALO-BERRY.

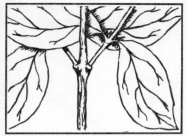

Fig. 165

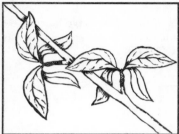

Fig. 167

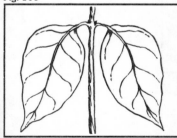

Fig. 166

Fig. 168

CAROLINA ALLSPICE
(Calycanthus floridus)
An aromatic shrub up to 9 ft. tall.
The flower is reddish-brown, 1 in.
long, with many stamens. The
fruit is a leathery capsule 1-2 in.
long. It grows in rich woods of the
Gulf Coast plain of Louisiana in
our area. It belongs to the
strawberry-shrub family
(Calycanthaceae).

THYME-LEAVED
MOCKORANGE
(Philadelphus serpyllifolius)
A shrub under 5 ft. tall with
straggly, brownish branches. The
flower is white, ½ in. across with
many stamens. The fruit is a
small, rounded capsule. It grows
on the lower mountain slopes in
New Mexico and western Texas.
It belongs to the saxifrage family
(Saxifragaceae).

BUTTONBUSH
(Cephalanthus occidentalis)
A shrub or tree up to 40 ft. tall.
The small white flowers are
borne in conspicuous globular
clusters about 1 in. in diameter.
The fruit is a small reddish nut. It
grows in low swampy ground and
in pond and stream margins
throughout all of our area. It
belongs to the madder family
(Rubiaceae).

CANADIAN
BUFFALO-BERRY
(Sheperdia canadensis)
A spreading shrub up to 8 ft. tall
with very small tubular flowers.
The fruit is berry-like, ⅓ in. long
and yellow or red. It grows on
open slopes at mid and high
altitudes in the New Mexico
mountains of our area. It belongs
to the oleaster family
(Eleagnaceae).

174

1. If the leaves have hairs on the veins of their underside, it is FRINGE-TREE.

2. If the leaves have black spots on their underside, it is COYOTILLO.

non-hairy

KI 88

3. If you found neither hairs nor black spots, proceed to the drawing (KI 89), at the right, in the left-hand margin of page 177.

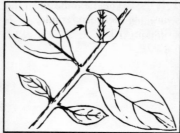

Fig. 169

FRINGE-TREE
(Chionanthus virginica)
A small tree up to 30 ft. tall with crooked, stout branches. The narrow, white petals of the flower are 1 in. long and twisted, bearing a purple spot at their base. The fruit is olive-shaped, bluish-black and about ¾ in. long. It grows in damp woods in eastern Oklahoma, Arkansas, Louisiana and eastern Texas. It belongs to the olive family *(Oleaceae)*.

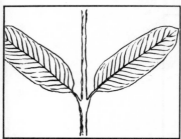

Fig. 170

COYOTILLO
(Karwinskia humboldtiana)
A shrub up to 6 ft. tall. The small flowers are greenish and few. The fruit is a rounded, conspicuous, brown to black olive (drupe), ⅓ in. in diameter and poisonous if eaten, causing muscular paralysis. It grows in dry prairies and plains in western and southwestern Texas, being common on the Rio Grande plain. It belongs to the buckthorn family *(Rhamnaceae)*.

neither-nor

KI 89

176

1. If the leaves of your specimen are narrowly elliptic and range in length from 2-6 in. it is WILD OLIVE.

2. If the leaves of your specimen are tear-shaped or ovate and ¾-2 in. long, it is ANISACANTH.

neither-nor

KI 89

3. If the leaves of your specimen are all ovate and ⅓-1½ in. long, it is SNOWBERRY.

4. If the leaves of your specimen are ovate or elliptical and range in length from 1½ to 4¾ in., it is an OISER.

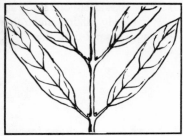

Fig. 171

Fig. 172

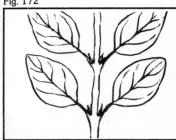

Fig. 173

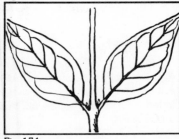

Fig. 174

WILD OLIVE
(Osmanthus americana)

A shrub or small tree up to 50 ft. tall with a narrow crown. Flowers imperfect and borne on separate trees, small and greenish-white. Fruits oval, olive-like, yellowish to purple and ½ to 1 in. in diameter. It is found in our area in southeastern Louisiana. It belongs to the olive family *(Oleaceae)*.

ANISACANTH
(Anisacanthus wrightii)

A low, irregular shrub up to 3 ft. tall. The slender, conspicuous flower is tubular, the tube curved and terminating in 2 lips, 1-2 in. long, and orange to red-colored. The fruit is a dry capsule about ¾ in. long. It grows on rocky stream banks in the Edwards Plateau region of Texas. It belongs to the acanthus family *(Acanthaceae)*.

SNOWBERRY
(Symphoricarpos spp.)

A group of shrubs up to 9 ft. tall with similar characteristics. If the fruits are white they are called SNOWBERRIES; if red they are called CORALBERRIES. The flower is bell- to funnel-shaped, pink or white. The fruit is a small, rotund drupe. They are common forest shrubs in our area. They belong to the honeysuckle family *(Caprifoliaceae)*.

OISER
(Cornus spp.)

A group of shrubs or small trees up to 15 ft. tall, separable mostly on floral cluster differences. Flowers minute, white. Fruit a rounded, dark-colored olive (drupe). They are found in seeps or along streams throughout our area. They belong to the dogwood family *(Cornaceae)*.

178

1. Check your specimen for the presence of naked thorns or thorn-tipped twigs, if present, go to the symbol (KI 91), shown to the right, in the left-hand margin of page 181.

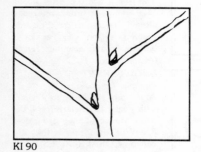

KI 90

2. If you found no thorniness in your specimen, look for the symbol (KI 97), shown to the right, in the left-hand margin of page 193.

KI 91

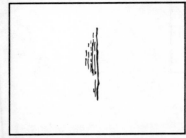

KI 97

1. If all of the leaves of your specimen have smooth (entire) blade margins, go to the symbol (KI 94), shown at the right, in the left-hand margin of page 187.

2. If some of the leaves are entire and some toothed, look for the drawing (KI 92), as shown at the right, in the left-hand margin of page 183.

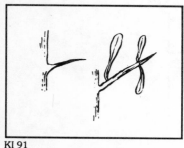

KI 91

3. If all of the leaves of your specimen have toothed, scalloped or lobed blade margins, go to the symbol (KI 93), shown to the right, in the left-hand margin of pages 185 and 185A.

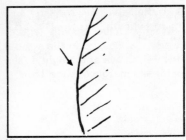

KI 94

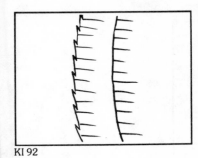

KI 92

KI 93

1. If the leaves are stalkless and hairy beneath, it is the DESERT PEACH.

2. If the leaves are stalked, their under surfaces lack hairs and their shape is oblong, as shown to the right, it is CRUCILLO.

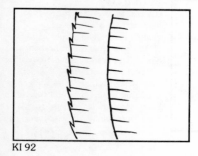

KI 92

3. If the leaves are stalked, their under surfaces hairless or slightly hairy and their outline is reverse oval or tear-shaped, as shown to the right, it is MEXICAN BLUEWOOD.

Fig. 175

DESERT PEACH
(Prunus fasiculata)
A thicket-forming shrub up to 8 ft. tall with dense, leafy branches. The lower is greenish-white, 1/3 in. across with many stamens. The fruit is a small, dry peach 3/5 in. long. It grows on dry, sunny hillsides in northern and western New Mexico. It belongs to the rose family *(Rosaceae).*

Fig. 176

CRUCILLO
(Condalia obtusifolia var. lycioides)
A rigid, spreading shrub up to 9 ft. tall with gray-green branches. The flower is small (1/8 in. across), flat and white. The fruit is a dark blue olive (drupe) 1/3 in. long. It grows on desert hills and flats of New Mexico. It belongs to the buckthorn family *(Rhamnaceae).*

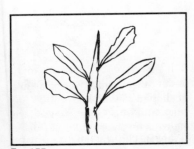
Fig. 177

MEXICAN BLUEWOOD
(Condalia spathulata var. mexicana)
A spreading shrub up to 10 ft. tall with rigid, gray-green branches. The flower is small and greenish and the fruit is a black olive (drupe) 1/4 in. long. It grows on desert hillsides in New Mexico. It belongs to the buckthorn family *(Rhamnaceae).*

184

1. If the leaves of your specimen are somewhat triangular in outline and lobed-toothed, as shown to the right, it is a LOBED-HAWTHORN.

2. If the blade margins are scalloped-sawtoothed, 2-5 in. long and densely hairy on their undersurface, it is WESTERN CRABAPPLE.

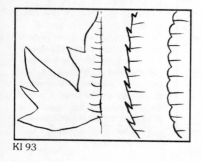

KI 93

3. If the blade margins are scalloped-sawtoothed, 1-3 in. long and the older leaves are hairless on their under surfaces, it is SOUTHERN CRABAPPLE.

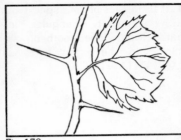
Fig. 178

LOBED-HAWTHORN
(Crataegus spp.)
Shrubs or small trees comprising a complex of species that intercross so that few species are distinct. The flower is normally flat, white and 5-petaled with many stamens. The fruit is a little apple (pome). It grows in woods and along streams throughout our area. It belongs to the rose family *(Rosaceae)*. Note: This is a "series species" within the genus.

Fig. 179

WESTERN CRABAPPLE
(Pyrus ioensis)
A tall shrub or small tree up to 30 ft. tall of irregular growth form with hairy twigs. The flower is ½ in. across, white or pink with 5 petals and many stamens. The fruit is a greenish-yellow apple about 1 in. in diameter. It grows in draws or along creeks in Arkansas, Louisiana, Oklahoma and Texas. It belongs to the rose family *(Rosaceae)*.

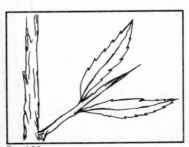
Fig. 180

SOUTHERN CRABAPPLE
(Pyrus angustifolia)
An open-headed, small tree up to 30 ft. tall. The showy flower is 1 in. across with 5 white to pink petals and many stamens. The fruit is a small, yellowish-green apple, 1 in. in diameter. It grows in woods and along river banks in southeast Texas, Arkansas, Oklahoma and Louisiana. It belongs to the rose family *(Rosaceae)*.

186

4. If the blade margins are saw-toothed, ½-1 in. long and hairless below, it is TEXAS BUCKTHORN.

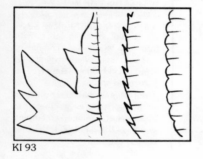

KI 93

5. If the margins are double saw-toothed, as shown to the right, it is SMOOTH HAW-THORN.

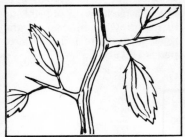

Fig. 181

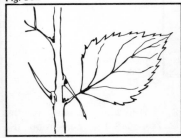

Fig. 182

TEXAS BUCKTHORN
(Condalia obtusifolia)
A stiff, gray-green branched shrub up to 8 ft. tall. The flower is small, green and inconspicuous. The fruit is a black olive (drupe) 2/5 in. in diameter. It is widespread in central and western Texas. It belongs to the buckthorn family *(Rhamnaceae)*.

SMOOTH HAWTHORN
(Crataegus spp.)
The thin-leaved series of haw intergraded species. See the notes about the *Crataegus* complex under LOBED-HAWTHORN at the top of this page.

1. Check for stalks on the leaves of your specimen. If present, go to the drawing (KI 95), at the right, in the left-hand margin of pages 189 and 189A.

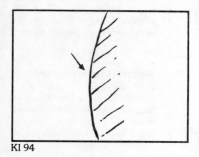

KI 94

2. If stalks are missing on the leaves, look for the symbol (KI 96), to the right, in the left-hand margin of page 191.

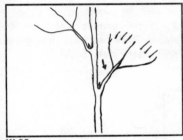

KI 95

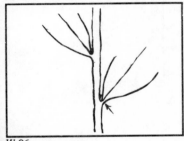

KI 96

1. If the leaves are oval to tear-shaped, 3-6 in. long and the older leaves are hairless, it is BOIS D'ARC.

2. If the leaves are oval to round, ⅜-¾ in. long and gray-scaly, it is SHADSCALE.

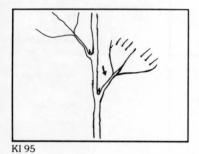

KI 95

3. If the leaves are elliptic in outline, ⅓-1 in. long, hairy on their upper surfaces and densely hairy below, it is BUCK-BRUSH.

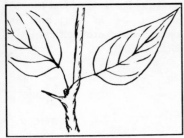

Fig. 183

BOIS D'ARC
(Maclura pomifera)
A tree up to 60 ft. tall with a rounded crown, arching older branches, milky sap and yellow wood. The flowers are imperfect, inconspicuous and borne on separate plants with the female tree bearing large (4-5 in. in diameter), multiple fruits that are wrinkled, green and milky. It is found mostly in waste ground, along fence rows or field edges in Arkansas, Oklahoma, Louisiana and the non-desert areas of Texas. It belongs to the mulberry family *(Moraceae)*.

Fig. 184

SHADSCALE
(Atriplex confertifolia)
A dense, clump-forming shrub up to 3 ft. tall. The flowers and fruits are difficult to discern. It grows on alkaline, desert soil in Trans-Pecos Texas and New Mexico. It belongs to the goosefoot family *(Chenopodiaceae)*.

Fig. 185

BUCK-BRUSH
(Ceanothus fendleri)
A low, thicket to mat-forming shrub up to 3 ft. tall. The small flowers occur in showy masses, the cluster about 1 in. long and white. The fruit is a tiny, brownish capsule. It grows in the mountains up to timberline in New Mexico and far western Texas. It belongs to the buckthorn family *(Rhamnaceae)*.

190

4. If the leaves are reverse tear-shape, 1/6-¾ in. long and hairless, it is GREEN BUCK-THORN.

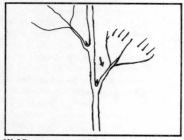

KI 95

5. If the leaves are reverse tear or egg-shape and up to 1½ in. long, it is BLUEWOOD.

Fig. 186

GREEN BUCKTHORN
(Condalia viridis)
A spreading shrub up to 9 ft. tall. Flowers inconspicuous; fruit black, olive-like and about 1/5 in. long. It grows in dry, limestone soils in the Edwards Plateau and Trans-Pecos regions of Texas. It belongs to the buckthorn family *(Rhamnaceae).*

Fig. 187

BLUEWOOD
(Condalia hookeri)
A thicket-forming shrub or small tree to 30 ft. tall. Small inconspicuous flowers without petals; the fruit a roundish, black olive (drupe) ⅓ in. in diameter. It is common on the Rio Grande plain and extends to central and western Texas. It belongs to the buckthorn family *(Rhamnaceae).*

1. If the leaves are narrow (linear), ½-1½ in. long and the older are hairless, it is GREASEWOOD.

2. If the leaves are reverse ovate to elliptic in outline, ⅜-1½ in. long, and some hairless and some slightly hairy, it is SPINY POLYGALA.

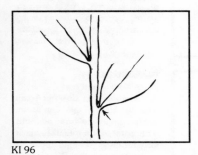

KI 96

3. If the leaves are reverse tear-shape, up to ½ in. long, some hairless and some hairy, it is GREASE-BUSH.

4. If the leaves are oblong-elliptic, ⅛-⅓ in. long, hairless but rough to the touch, it is MORTONIA.

191

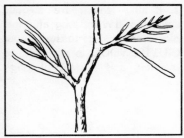

Fig. 188

GREASEWOOD
(Sarcobatus vermiculatus)
A many-branched, rigid shrub up to 10 ft. tall with whitish young branches, becoming gray with age. The flowers and fruits are difficult to discern, although they are borne in long spikes with red-tinged bracts. It grows on the desert flats of New Mexico and northwestern Texas. It belongs to the goosefoot family *(Chenopodiaceae).*

Fig. 189

SPINY POLYGALA
(Polygala subspinosa)
A very low (1 ft. tall) shrub with erratic grayish-green, slender branches. Flower tubular, irregular in shape and yellow to purple in color. Fruit a dry capsule ¼ in. long. It is found on dry hillsides and plains in New Mexico. It belongs to the milkwort family *(Polygalaceae).*

Fig. 190

GREASE-BUSH
(Forsellesia spinescens)
A densely, angular-branched shrub up to 5 ft. tall. The flower is inconspicuous, small and white. The fruit is dry, leathery and about 1/5 in. long. It grows on rocky limestone hills in New Mexico and western Texas. It belongs to the staff-tree family *(Celastraceae).*

Fig. 191

MORTONIA
(Mortonia scabrella)
An erect, yellowish-green shrub up to 8 ft. tall with stiff, very leafy branches. The flower is 1/6 in. across and white. The fruit is a dry capsule ca. ¼ in. long. It grows on dry hillsides in New Mexico and western Texas. It belongs to the staff-tree family *(Celastraceae).*

1. If all of the leaves of your specimen have smooth (entire) blade margins, go to the symbol (KI 112), shown to the right, in the left-hand margin of page 223.

2. If *some* of the leaves are entire and *some* toothed or lobed, go to the drawing (KI 98), at the right, and in the left-hand margin of page 195.

KI 97

3. If *all* of the leaves of your specimen have toothed, scalloped or lobed margins, go to this symbol (KI 138), at the right, in the left-hand margin of page 275.

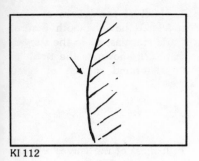

KI 112

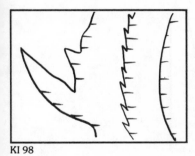

KI 98

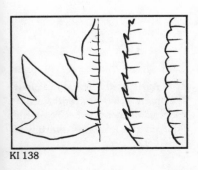

KI 138

1. If most of the leaves are apparently stalkless, your specimen is either BLUERIDGE (½-3 ft. tall) or the ELLIOTT (3-12 ft. tall) BLUEBERRY.

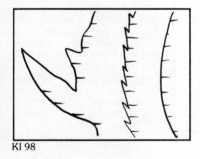

KI 98

2. If most of the leaves of your specimen are apparently stalked, find drawing (KI 99), at the right, in the left-hand margin of page 197.

Fig. 192

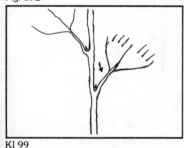

KI 99

BLUERIDGE OR ELLIOTT BLUEBERRY

(Vaccinium vacillans or *V. elliottii)*
The blueridge is a low shrub (3 ft. tall) and the elliott blueberry is a taller shrub up to 12 ft. tall. In both, the flower is ⅓ in. long and urn-shaped. The fruit is a dark blue berry ⅓ in. in diameter. The blueridge blueberry grows in old fields and dry woods in Arkansas and the elliott blueberry grows on low ground in Arkansas, Louisiana and eastern Texas. They belong to the heath family *(Ericaceae).*

1. Look on both sides of the leaves of your specimen, if hairless, follow the drawing (KI 100), as shown to the right, to the left margin of page 199.

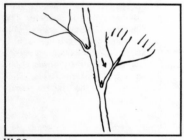

KI 99

2. If either side of the leaves is hairy, look for the diagram (KI 105), shown to the right, in the left-hand margin of page 209.

197

non-hairy

KI 100

hairy

KI 105

198

1. If the leaves of your specimen are tear-shaped, as shown to the right, (KI 101), go to the left margin of page 201.

non-hairy

KI 100

2. If the leaves of your specimen are reverse tear-shape, ovate, elliptic or reversed ovate-elliptic, find the symbol (KI 102), shown to the right, in the left-hand margin of page 203.

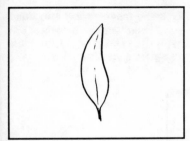

KI 101

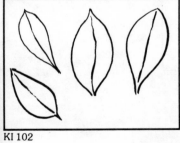

KI 102

1. If the leaves are 2-4 in. long and their undersurfaces are more white than green, it is BLUESTEM WILLOW.

2. If the leaves are 2-4 in. long and their undersurfaces are green, it is CAROLINA CHERRY-LAUREL.

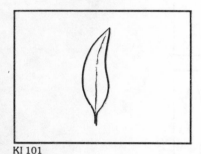

KI 101

3. If the leaves are 1½-3 in. long, their undersurfaces are green and their young branches have distinctive red, smooth bark, it is ARIZONA MA-DRONE.

Fig. 193

BLUESTEM WILLOW
(Salix irrorata)
A shrub up to 12 ft. tall with brownish stems covered with a bluish bloom. Flowers imperfect, on separate trees and borne in pendulous spikes. The fruit is a tiny capsule. It grows along streams in the mountains of New Mexico. It belongs to the willow family *(Salicaceae)*.

Fig. 194

CAROLINA CHERRY-LAUREL
(Prunus caroliniana)
An evergreen tree up to 35 ft. tall. The small, white flowers occur in elongate clusters. The fruit is a black olive (drupe) ½ in. long. It grows in lowland forests in south and central and eastern Texas across into Louisiana. It belongs to the rose family *(Rosaceae)*.

Fig. 195

ARIZONA MADRONE
(Arbutus arizonica)
An evergreen tree up to 40 ft. tall with smooth-barked, red branches. The flower is small and white to pink. The fruit is berry-like, orange to red and ⅓ in. in diameter. It grows on exposed slopes in the San Luis and Animas mountains of New Mexico. It belongs to the heath family *(Ericaceae)*.

1. If the leaves of your specimen are reverse egg- or tear-shape, as shown to the right in (KI 103), go to the left-hand margin of page 205.

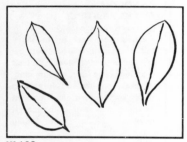

KI 102

2. If the leaves of your specimen are egg or elliptic in outline, find the symbol (KI 104), shown to the right, in the left-hand margin of pages 207 and 207A.

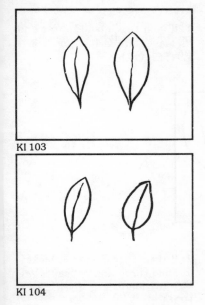

KI 103

KI 104

1. If your specimen is a tree and its leaves are 2-4 in. long, it is WATER OAK.

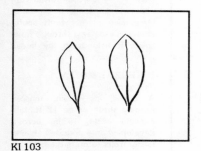

KI 103

2. If your specimen is a thicket-forming shrub and its leaves are ⅝-2 in. long, it is INKBERRY HOLLY.

Fig. 196

WATER OAK
(Quercus nigra)
A tree up to 45 ft. tall with mostly smooth, black bark. The tiny flowers are borne in pendulous spikes. The fruit is an acorn (nut enclosed by a woody, basal cup) about ⅔ in. long. It grows in streamside forests from south-central Texas east through Louisiana. It belongs to the beech family *(Fagaceae)*.

Fig. 197

INKBERRY HOLLY
(Ilex glabra)
A dense, evergreen, thicket-forming shrub up to 12 ft. tall. Flowers white, ¼ in. across, developing into rounded, black, fleshy fruits ⅓ in. in diameter. It grows near the coast in sandy, acid bogs of Louisiana and is rare in northeast Texas. It belongs to the holly family *(Aquifoliaceae)*.

1. If the stalks of the *older* leaves are hairy and the leaves are ½-2 in. long, it is MYRTLE OAK.

2. If the stalks of the *older* leaves are non-hairy, longer than 2 in. but aromatic when crushed in your hand, it is SASSAFRAS.

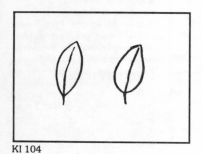

KI 104

3. If the older leaf stalks are non-hairy, the leaves longer than 2 in., not aromatic and only the young leaves sharp-toothed in the manner shown to the right, it is LAUREL OAK.

Fig. 198

MYRTLE OAK
(Quercus hemisphaerica)
A shrub or small tree up to 30 ft. tall. Flowers inconspicuous, borne in pendent spikes. Fruit a shiny, dark brown acorn, ½ in. long. It grows in sandy woodlands and prairies along the coasts of Texas and Louisiana. It belongs to the beech family *(Fagaceae)*.

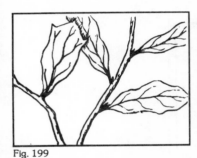

Fig. 199

SASSAFRAS
(Sassafras albidum)
A tree up to 100 ft. tall with aromatic bark and short, crooked branches. Flowers imperfect, sexes separate, petals absent. Fruit an oval, blue drupe, ½ in. long. It grows in abandoned, cultivated fields, sandy woods and fence rows in eastern Texas, Oklahoma, Arkansas and Louisiana. It belongs to the laurel family *(Lauraceae)*.

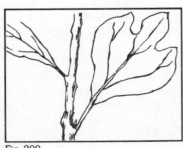

Fig. 200

LAUREL OAK
(Quercus laurifolia)
A large tree up to 70 ft. tall with blackened bark and a dense crown. Flowers inconspicuous and borne in pendulous spikes. Fruit is a dark brown acorn, ½ in. long. It grows in low ground in southeastern Texas and southern Louisiana. It belongs to the beech family *(Fagaceae)*.

4. If the older leaf stalks are non-hairy, the leaves 2-5 in. long, not aromatic and only older leaves irregularly toothed in the manner shown to the right, it is LIVE OAK.

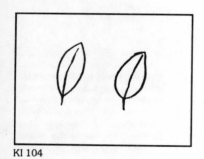

KI 104

5. If the older leaf stalks are non-hairy, the leaves are 1-3 in. long, non-aromatic and toothed leaves regularly toothed, it is CALIFORNIA BUCKTHORN.

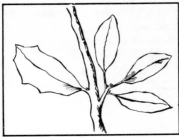

Fig. 201

LIVE OAK
(Quercus virginiana)
An evergreen tree with spreading branches low to the ground, up to 60 ft. tall. Flowers small, borne in pendent spikes. Fruit a blackish acorn about ½ in. long. It occurs in eastern Oklahoma and central Texas east into Louisiana, particularly along the coast. It belongs to the beech family *(Fagaceae)*.

Fig. 202

CALIFORNIA BUCKTHORN
(Rhamnus californica)
Mostly a shrub, rarely a tree up to 18 ft. tall. The flower is inconspicuous and greenish. The fruit is berry-like, green to red and ¼ in. in diameter. It grows in sheltered canyons in the midmontane regions of New Mexico. It belongs to the buckthorn family *(Rhamnaceae)*.

1. If the leaves of your specimen are prominently or only hairy on the undersurface, go to the drawing (KI 106), shown to the right, in the left-hand margin of page 211.

2. If the leaves are prominently hairy on both upper and lower surfaces and the young twigs, it is SILVER-LEAF WILLOW.

KI 105

3. If the leaves are prominently hairy on both surfaces but the twigs are non-hairy, it is BEAKED WILLOW.

hairy

KI 106

Fig. 203

SILVER-LEAF WILLOW
(Salix argophylla)
A thicket-forming shrub or small tree up to 18 ft. tall. The inconspicuous flowers are borne in pendulous spikes and the fruit is a tiny, silky capsule. It grows along mid-mountain streams in Trans-Pecos Texas and New Mexico. It belongs to the willow family *(Salicaceae)*.

Fig. 204

BEAKED WILLOW
(Salix bebbiana)
A shrub or small tree up to 20 ft. tall. Flowers occur in lax spikes. The fruit is a silky capsule. It grows along most mountain streams in New Mexico. It belongs to the willow family *(Salicaceae)*.

210

1. If the leaves of your specimen are reverse egg-shape, as shown to the right, it is DURAND OAK.

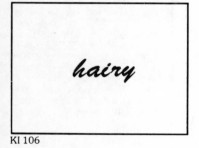

KI 106

2. If the leaves of your specimen are ovate, elliptic or tear-shaped, go to (KI 107), at the right, in the left-hand margin of page 213.

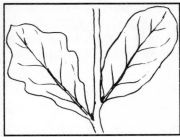

Fig. 205

DURAND OAK

(Quercus sinuata)

A large, dense-crowned tree up to 100 ft. tall. The inconspicuous flowers are borne in lax spikes. The fruit is an acorn (nut with basal, woody cup) around ½ in. long. Found along rivers in central Texas and the forests of eastern Texas, southeastern Oklahoma, Arkansas and Louisiana. It belongs to the beech family *(Fagaceae)*.

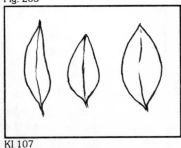

KI 107

1. Look at the leaf stalks of the *older* leaves of your specimen. If hairy, proceed to (KI 108), at the right, in the left-hand margin of page 215.

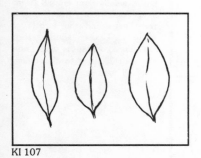

KI 107

2. If the leaf stalks of the *older* leaves, are non-hairy, go to the symbol, (KI 110), shown to the right, in the left-hand margin of page 219.

hairy

KI 108

non-hairy

KI 110

214

1. If the upper surfaces of the leaves of your specimen are rough-bumped, as shown to the right, it is ANACUA.

2. If the upper surface is smooth, but the twigs are hairy, it is GRAY OAK.

hairy

KI 108

3. If the upper surface is smooth, the twigs non-hairy, but the leaf stalks are grooved and the leaves are 5-10 in. long, it is SWAMP TUPELO.

4. If the upper surface is neither bumpy, nor the twigs hairy, nor the leaf stalks grooved and leaf length is less than 5 in., go to (KI 109), at the right, in the left-hand margin of page 217.

Fig. 206

ANACUA
(Ehretia anacua)

A shrub or several-trunked tree up to 45 ft. tall with reddish-scaled bark and evergreen leaves in the southern part of its range. The fragrant, white flowers are ¼ in. long but occur in 3 in. long clusters. The fruit is a yellow-orange cherry (drupe) ⅓ in. in diameter. It grows in upland thickets and along streams in southern Texas. It belongs to the borage family *(Boraginaceae)*.

Fig. 207

GRAY OAK
(Quercus grisea)

A large shrub or tree up to 50 ft. tall. The flowers are minute and borne in flexulous spikes. The fruit is an acorn (nut with basal cup) ¾ in. long. It grows on exposed, rocky slopes in the mid-altitudes of western Texas and southern New Mexico mountains. It belongs to the beech family *(Fagaceae)*.

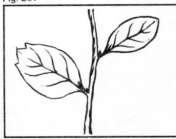

Fig. 208

SWAMP TUPELO
(Nyssa aquatica)

A large tree up to 90 ft. tall with a trunk swollen at its base. Flower small, fleshy, developing into an oval, purple olive about 1 in. long. It grows in swamps and sluggish streams in eastern Oklahoma, eastern Texas, Arkansas and Louisiana. It belongs to the dogwood family *(Cornaceae)*.

neither-nor

1. If you have found your specimen in Southwestern New Mexico and its entire leaves look like the symbol to the right, it is TOUMEY OAK.

2. If you have found your specimen in Southwestern Oklahoma or North-central Texas and its entire leaves look like the symbol to the right, it is LIMESTONE OAK.

neither-nor

KI 109

3. If you have found your specimen along the coast in Southern Louisiana and its entire leaves look like the symbol to the right, it is MYRTLE OAK.

217

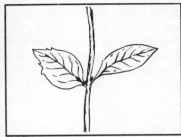
Fig. 209

TOUMEY OAK
(Quercus toumeryi)
A shrub about 6 ft. tall or rarely a tree up to 25 ft. tall. Flowers inconspicuous in lax spikes; fruit an acorn about ⅔ in. long. It grows at mid-altitudes on open slopes in the mountains of southwestern New Mexico. It belongs to the beech family *(Fagaceae)*.

Fig. 210

LIMESTONE OAK
(Quercus mohriana)
A thicket-forming shrub or sometimes a small tree up to 18 ft. tall. Flowers inconspicuous in lax spikes, the fruit an acorn 1/16 in. long. It grows in dry, limey soils in west-central and western Texas and southwestern Oklahoma. It belongs to the beech family *(Fagaceae)*.

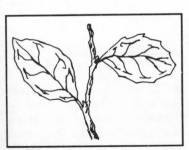
Fig. 211

MYRTLE OAK
(Quercus hemisphaerica)
A shrub or small tree up to 30 ft. tall. Flowers inconspicuous in pendent spikes. Fruit a dark brown, shiny acorn ½ in. long. It is found in sandy woodlands and prairies along the coasts of Texas and Louisiana. It belongs to the beech family *(Fagaceae)*.

218

1. If the older leaves of your specimen possess leaf-stalk outgrowths, as shown to the right, it is PRAIRIE WILLOW.

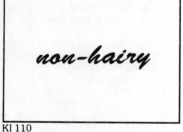

non-hairy

KI 110

2. If the *older* leaf stalks lack such "stipules", to to (KI 111), at the right, in the left-hand margin of page 221.

Fig. 212

PRAIRIE WILLOW
(Salix humilis)
A shrub up to 9 ft. tall with clustered, wand-like stems. The flowers are borne in lax spikes and the fruit is a small capsule. It grows in open areas near water in Arkansas, Oklahoma, eastern Texas and Louisiana. It belongs to the willow family *(Salicaceae)*.

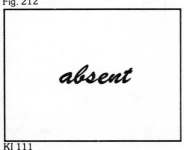

absent

KI 111

1. If your specimen occurs in bogs above timberline in the mountains of New Mexico, it is the PLANE-LEAVED WILLOW.

2. If your specimen occurs below timberline in the mountains of New Mexico, it is SCOULAR WILLOW.

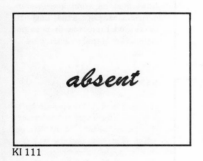

KI 111

3. If your specimen occurs in rich, moist soil in Arkansas, east Texas, east Oklahoma, and Louisiana and its toothed leaves are regularly toothed as shown to the right, it is STYRAX.

4. If your specimen occurs in Texas, Oklahoma and Louisiana and its toothed leaves are irregularly toothed, as shown to the right, it is LIVE OAK.

221

Fig. 213

PLANE-LEAVED WILLOW
(Salix planifolia)
A many-branched shrub up to 10 ft. tall with flowers borne in lax spikes, the fruit a small capsule. It grows near or above timberline in the high mountains of New Mexico in our area. It belongs to the willow family *(Salicaceae)*.

Fig. 214

SCOULAR WILLOW
(Salix scouleriana)
A thicket-forming shrub or tree up to 30 ft. tall with slender, pendulous branches. Flowers occur in dense spikes and the fruit is a small capsule. It grows on mountain slopes up to 11,000 ft. in New Mexico. It belongs to the willow family *(Salicaceae)*.

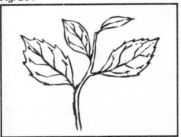
Fig. 215

STYRAX
(Styrax spp.)
A group of shrubby species up to 9 ft. tall. The fragrant flowers are drooping, white, conspicuous, about ¼ in. long and the fruit is a dry capsule about ⅓ in. in diameter. They occur in moist, rich habitats in the eastern portion of our area. They belong to the storax family *(Styracaceae)*.

LIVE OAK
(Quercus virginiana)
See page 208A

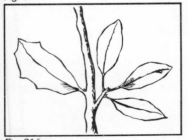
Fig. 216

1. If the leaf lacks a stalk, as shown to the right, go to (KI 113), in the left-hand margin of page 225.

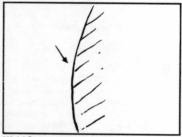

KI 112

2. If the leaf possesses a stalk, as shown to the right, in (KI 117), look for this picture in the left-hand margin of page 233.

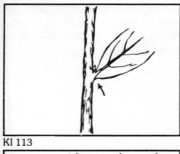

KI 113

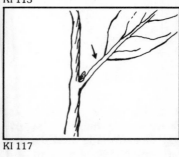

KI 117

1. If the blade is narrow in relation to its length and the margins parallel (linear), go to (KI 114), shown to the right, in the left-hand margin of page 227.

KI 113

2. If the blade is tear-shape, elliptical, egg-shape or reverse egg- or tear-shape, find the picture, (KI 115), shown to the right, in the left-hand margin of pages 229 and 229A.

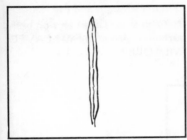

KI 114

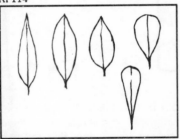

KI 115

1. If the leaves are sticky to the touch (viscid), it is RABBIT-BRUSH.

2. If the leaves are hairy on both surfaces, it is YEW-LEAVED WILLOW.

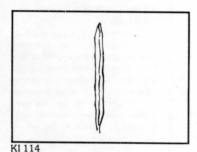

KI 114

3. If the leaves are gray-dandruffed on both sides, it is WINGED SALTBUSH.

Fig. 217

RABBIT-BRUSH
(Chrysothamnus spp.)
A group of many-branched shrubs from 2 to 10 ft. tall. Flower clusters are yellow and aster-like with small, seed-like fruits. They grow on the high plains of Texas and Oklahoma west into New Mexico. They belong to the aster family *(Compositae)*.

Fig. 218

YEW-LEAVED WILLOW
(Salix taxifolia)
A short-branched, large shrub or tree up to 50 ft. tall. Flowers tiny, in rounded spikes that are ½ in. long, the fruit being a small capsule. This willow grows along stream banks at mid-altitudes of the mountains in far western Texas and southwestern New Mexico. It belongs to the willow family *(Salicaceae)*.

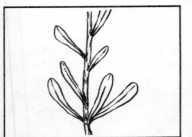

Fig. 219

WINGED SALTBUSH
(Atriplex canescens)
A many-branched evergreen shrub up to 8 ft. tall. Flowers inconspicuous in dense, leafy spikes, the spike leaves distinctively 4-winged; fruit tiny and seed-like. This shrub is quite common on the prairies and desert flats of New Mexico, western Texas and western Oklahoma. It belongs to the goosefoot family *(Chenopodiaceae)*.

228

1. If the leaves of your specimen are white-hairy on both sides and tear-shaped or egg-shaped, it is THIN-BRUSH.

2. If the leaves of your specimen are silvery-hairy on both surfaces and elliptic in shape, it is ARROW-WEED.

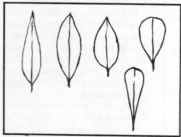

KI 115

3. If the lower surface is hairy on some leaves and not on others and the blade outline is egg-shape (ovate) or reverse ovate, your specimen is LEATHER-WOOD.

Fig. 220

THIN-BRUSH
(Dicraurus leptocladus)
A low silvery-green shrub 2-3 ft. tall. Flowers tiny borne in hairy, terminal clusters; fruit tiny, seed-like. It grows on exposed rocky slopes in the southern Trans-Pecos area of Texas. It belongs to the pigweed family *(Amaranthaceae)*.

Fig. 221

ARROW-WEED
(Tessaria sericea)
A "rank"-odored shrub with many erect, willow-like, gray-green branches mostly around 3 ft. tall. The flowers are grouped into many small pink heads. The fruits are small, seed-like. It grows near streams in New Mexico and Trans-Pecos Texas. It belongs to the aster family *(Compositae)*.

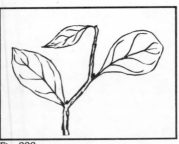

Fig. 222

LEATHERWOOD
(Dirca palustris)
A gray, smooth-barked shrub up to 7 ft. tall. Flower yellowish, funnel-form, ½ in. long. Fruit red to orange, olive-like and 2/5 in. long. It grows in rich, wet areas in Louisiana forests. It belongs to the mezerum family *(Thymelaeaceae)*.

230

4. If the lower surface is hairy and glandular and the blade elliptic in shape, ¾-1½ in. long, it is BUSH HUCKLEBERRY.

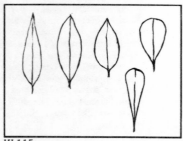

KI 115

5. If the leaves are non-hairy on both surfaces, (KI 116), to the right, find this drawing in the left-hand margin of page 231.

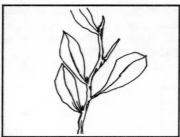

Fig. 223

BUSH HUCKLEBERRY
(Gaylussacia dumosa)
A glandular-hairy shrub with erect stems from 1 to 2 ft. tall. The white to pink bell-shaped flowers are about ¼ in. long and borne in dense clusters. The fruit is a black berry about ¼ in. in diameter. It grows in acid swamps and soils in Louisiana. It belongs to the heath family *(Ericaceae)*.

non-hairy

KI 116

230A

1. Look at the blade outline of the leaves of your specimen. If it is ovate or elliptical, it is NORTHERN ANDRACHNE.

2. If the leaf blade of your specimen is reverse egg-shape, as shown to the right, it is DESERT YAUPON.

non-hairy

KI 116

3. If the leaf blade is narrow and the margins parallel (linear), and the surfaces sticky to the touch, it is ROSIN-BUSH.

4. If the leaves are linear and the surfaces non-sticky to the touch, it is RABBIT-BRUSH.

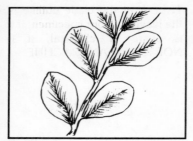

Fig. 224

NORTHERN ANDRACHNE
(Andrachne phyllanthoides)
A diffusely branched shrub up to 3 ft. tall. Flowers minute, greenish; the fruit a small, dry capsule. It is uncommon on limestone outcrops near streams in the Edwards Plateau and north-central areas of Texas, north into Oklahoma and east into Arkansas. It belongs to the spurge family *(Euphorbiaceae)*.

Fig. 225

DESERT YOUPON
(Schaefferia cuneifolia)
A rigidly branched shrub from 3 to 6 ft. tall with light-gray bark. The flowers minute, greenish. The fruit olive-like, orange to red, minute. It grows on rocky slopes and in canyons in southwestern Texas. It belongs to the staff-tree or bittersweet family *(Celastraceae)*.

Fig. 226

ROSIN-BUSH
(Baccharis sarothroides)
A few-leaved, broom-like, evergreen shrub up to 9 ft. tall. Flowers small in small clusters, the fruit small and seed-like. It is found in desert grasslands or sandy washes in New Mexico. It belongs to the aster family *(Compositae)*.

Fig. 227

RABBIT-BRUSH
(Chrysothamnus spp.)
A group of many-branched shrubs from 2 to 10 ft. tall. Aster-like flowers yellow, the fruit small, seed-like. They grow in the high plains of Texas and Oklahoma and west into New Mexico. They belong to the aster family *(Compositae)*.

1. If the blade is neither hairy nor scaly (glabrous), find this drawing (KI 118), to the right, in the left-hand margin of pages 235 and 235A.

KI 117

2. If either the upper or the lower surface of the blade or just its veins are hairy or scaly, find the symbol (KI 125), to the right, in the left-hand margin of page 249.

neither-nor

KI 118

hairy

KI 125

1. If the blade outline is diamond-shaped, it is DIAMOND-LEAF OAK.

2. If the blade is heart-shaped, as shown to the right, with a sharp tip, it is REDBUD.

neither-nor

KI 118

3. If the blade is heart-shaped but the tip rounded, it is CALIFORNIA REDBUD.

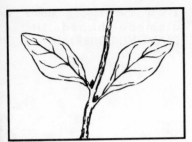

Fig. 228

DIAMOND-LEAF OAK
(Quercus laurifolia var. *obtusa)*
See LAUREL OAK, page 208
for a description of this tree.

Fig. 229

REDBUD
(Cercis canadensis)
A shrub or small tree up to 40 ft.
tall with a broad-rounded to flattened crown. Flower appearing
before the leaves, rose-purple,
and pea-like. Fruit a pod 2 to 4
in. long and about ½ in. wide. It
occurs in streamside forests in
Oklahoma, Arkansas, Louisiana
and the eastern third of Texas. It
belongs to the pea family
(Leguminosae).

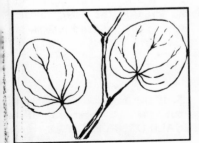

Fig. 230

CALIFORNIA REDBUD
(Cercis occidentalis)
Mostly a shrub or sometimes a
small tree up to 20 ft. high with
smooth, gray bark. For a description of flowers and fruit, see the
notes for the REDBUD above. It
grows on moist slopes or along
streams in New Mexico for our
area. It belongs to the pea family
(Leguminosae).

4. If the blade is curved tear-shaped (falcate), as shown to the right, it is SUGAR HACK-BERRY.

neither-nor

KI 118

5. If the blade shape is oval, elliptic, linear tear-shape and reverse oval or tear-shape, find this drawing (KI 119), to the right, in the left-hand margin of page 237.

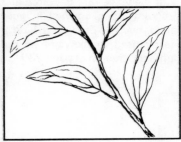

Fig. 231

SUGAR HACKBERRY
(Celtis laevigata)
A tree up to 90 ft. tall, its thin bark bearing irregular, corky outgrowths. Flower inconspicuous, the fruit globose, orange-red to black and ¼ in. in diameter. It occurs mostly along streams in Texas, Oklahoma, Arkansas and Louisiana in our area. It belongs to the elm family *(Ulmaceae).*

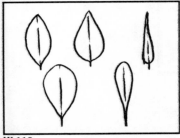

KI 119

236A

1. If the lower surface is black-dotted, it is FETTERBUSH.

•

2. If the leaves are undotted and fragrant when crushed in your hand, find this drawing (KI 120), as pictured to the right, in the left-hand margin of page 239.

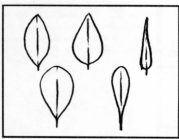

KI 119

3. If the leaves are neither dotted nor fragrant when crushed, find their symbol (KI 121), at the right, in the left-hand margin of pages 241 and 241A.

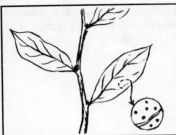

Fig. 232

FETTERBUSH
(Lyonia lucida)
A drooping shrub 2-6 ft. tall. The elongate, urn-shaped flowers are white to pink and about ⅓ in. long. The fruit is a lobed, rounded, small capsule. It grows in low, wet sandy soils in southeastern Louisiana in our 5-state area. It belongs to the heath family *(Ericaceae)*.

fragrant

KI 120

neither-nor

KI 121

238

1. If the leaves are reverse ovate and 4-11 in. long, it is PAWPAW.

2. If the leaves are ⅔-2 in. long with hairy leaf stalks, it is TAR-BUSH.

fragrant

KI 120

3. If the leaves are elliptic, 2-5 in. long and your specimen a several-stemmed shrub, it is SPICEBUSH.

4. If the leaves are elliptic, 3-4 in. long and your specimen is a single-stemmed tree, it is RED BAY.

239

Fig. 233

PAWPAW
(Asimina triloba)
A shrub or tree up to 35 ft. tall. Flower purplish-green, 1-2 in. long; fruit pulpy, aromatic 2-7 in. long. It occurs in rich, bottomlands in Arkansas, Louisiana and eastern Texas. It belongs to the custard-apple family *(Annonaceae)*.

Fig. 234

TARBUSH
(Flourensia cernua)
A very leafy shrub 3-6 ft. tall with a tar-like odor. Flowers small, in yellowish heads ⅓ in. long. Fruit small, seed-like. It is very common in the deserts of Trans-Pecos Texas and New Mexico. It belongs to the aster family *(Compositae)*.

Fig. 235

SPICEBUSH
(Lindera benzoin)
A shrub or small tree with aromatic foliage. Flowers yellow, fragrant, ⅓ in. across, appearing before the leaves. Fruit cherry-like, red, spicy, 2/5 in. in diameter. It grows in swampy soil in eastern Oklahoma, eastern Texas, Arkansas and Louisiana. It belongs to the laurel family *(Lauraceae)*.

RED BAY
(Persea borbonia)
An evergreen tree up to 60 ft. tall. Flowers few, small, yellowish; fruit olive-like, blue to black, ½ in. long. It grows in damp woods, swamps and long streams in Arkansas, Oklahoma, but mostly near the coast in southern Louisiana and southeastern Texas. It belongs to the laurel family *(Lauraceae)*.

Fig. 236

240

1. If the blade margins are parallel and the blade narrow, it is ROOSEVELT WEED.

neither-nor

KI 121

2. If the blade outline is reverse egg-shape, it is ARKANSAS OAK.

3. If the blade outline is reverse tear-shape and ½-1¼ in. long, it is BEARBERRY.

Fig. 237

ROOSEVELT WEED
(Baccharis neglecta)
A shrub 3-9 ft. tall with striate-angled branchlets. Flowers and fruits inconspicuous. It grows in marshes in Louisiana and eastern Texas. It is in the aster family *(Compositae)*.

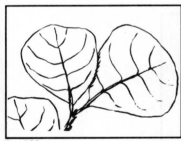
Fig. 238

ARKANSAS OAK
(Quercus arkansana)
A tree up to 60 ft. tall. Flowers inconspicuous; fruit an acorn, ⅓ in. long. It grows in sterile clays and sands in Arkansas. It is in the beech family *(Fagaceae)*.

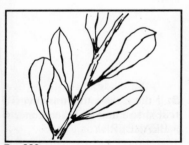
Fig. 239

BEARBERRY
(Arctostaphylos uva-ursi)
A prostrate, evergreen shrub up to 1 ft. high. Flowers ¼ in. long, white-pink, urn-shaped; fruit ovoid, bright red, ⅓ in. in diameter. Common in the montane pine and spruce belts of New Mexico. Of the heath family *(Ericaceae)*.

4. If the blade is reverse tear-shaped, 1-4 in. long, and the young twigs hairy, it is DA-HOON HOLLY.

5. If the blade is reverse tear-shape, 2-4 in. long and the young twigs smooth, it is CYRILLA.

KI 121

6. If the blade is elliptic or oval in outline, find this drawing (KI 122), to the right, in the left-hand margin of page 243.

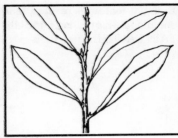

Fig. 240

DAHOON HOLLY
(Ilex cassine)

An evergreen shrub or tree up to 30 ft. tall. The flowers are white and 1/6 in. across. The fruit is globose, red to orange and conspicuous in massed clusters, each ¼ in. in diameter. It is a plant of swampy ground in southeastern Texas and Louisiana. It belongs to the holly family *(Aquifoliaceae)*.

Fig. 241

CYRILLA
(Cyrilla racemiflora)

Shrub or small tree up to 25 ft. tall. Flowers tiny, white, borne in 6 in. long spikes. Fruit a tiny capsule. It grows in swamps in southern Louisiana and southeastern Texas. It belongs to the cyrilla family *(Cyrillaceae)*.

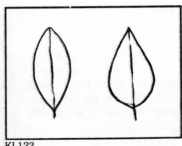

KI 122

1. If the leaf stalk is hairy and the blade elliptic, it is MOUNTAIN LAUREL.

2. If the leaf stalk is smooth and the leaves are less than ¾ in. long, it is EVERGREEN BLUEBERRY.

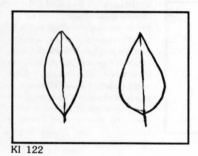

KI 122

3. If the leaf stalk is smooth and some of the leaves exceed ¾ in. in length, go to this drawing, (KI 123), to the right, in the left-hand margin of pages 245 and 245A.

Fig. 242

MOUNTAIN LAUREL
(Kalmia latifolia)
A thicket-forming, evergreen shrub or tree up to 30 ft. Flowers showy, saucer-shaped, white or pink, 1 in. across. Fruit a dry capsule, ¼ in. in diameter. It grows in woods or along swamp margins in eastern Louisiana. It belongs to the heath family *(Ericaceae)*.

Fig. 243

EVERGREEN BLUEBERRY
(Vaccinium darrowii)
An evergreen shrub growing in dense colonies that are seldom over 16 in. high. The small (¼ in. long) flowers are red or pink and urn-shaped. The fruit is a blue berry about ¼ in. in diameter. It grows in sandy soil in open areas along streams in southern Louisiana and southeastern Texas. It belongs to the heath family *(Ericaceae)*.

KI 123

1. If your specimen grows above timberline in the mountains of New Mexico to a height of 6 in., it is ROCKY MOUNTAIN WILLOW.

2. If you have found your specimen on the beaches of Texas with some of its stems growing as runners, it is MAYTEN.

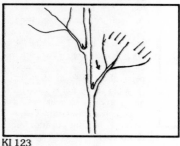

KI 123

3. If your specimen is an upright shrub found in the mountains of New Mexico with its leaves 3-veined from the base, as shown to the right, it is CHAGUIRA.

Fig. 244

ROCKY MOUNTAIN WILLOW

(Salix saximontana)
Prostrate shrub scarcely 6 in. tall. Tiny flowers are borne in short spikes. Fruit a tiny capsule. It grows above timberline in New Mexico. It is in the willow family *(Salicaceae)*.

Fig. 245

MAYTEN

(Maytenus texana)
A prostrate, thicket-forming shrub. Flower greenish-white, ⅛ in. across. Fruit an ovoid, small capsule. It grows on clay mounds along the coast from Corpus Christi to Brownsville, Texas. It is in the bittersweet family *(Celastraceae)*.

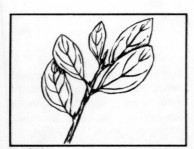

Fig. 246

CHAGUIRA

(Ceanothus integerrimus)
A shrub growing 2-10 ft. tall. The tiny white flowers occur in large showy masses. The fruit is a roundish capsule ¼ in. in diameter. It grows on arid sites in the lower reaches of the Mogollon mountains in New Mexico. It belongs to the buckthorn family *(Rhamnaceae)*.

4. If your specimen is a thicket-forming shrub found in the mountains of Texas and New Mexico with its leaves several-veined from the midrib, it is MANZANITA.

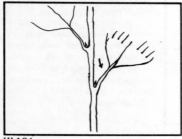

KI 124

5. If your specimen grows in low, acid, sandy soil in Arkansas, East Texas and Louisiana, go to the drawing, (KI 124), to the right, in the left-hand margin of page 247.

Fig. 247

MANZANITA
(Arctostaphylos pungens)
An erect, thicket-forming shrub up to 9 ft. tall with smooth, red bark. Flower urn-shaped, small, white or pink. Fruit berry-like, red, ⅓ in. in diameter. It grows on dry mountain slopes in Trans-Pecos Texas and New Mexico. It is in the heath family *(Ericaceae)*.

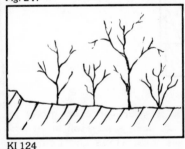
KI 124

1. If the leaves have a whitened undersurface, it is SWAMP MAGNOLIA.

2. If the leaf undersurface is green, the young twigs glandular-sticky and plant height between 1-4 ft., it is BLACK HUCKLEBERRY.

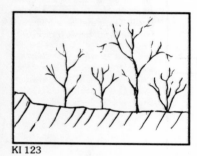

KI 123

3. If the leaf undersurface is green, the young twigs not sticky to the touch and the plant height between 3-12 ft., it is CANDLEBERRY.

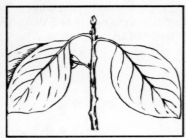

Fig. 248

SWAMP MAGNOLIA
(Magnolia virginiana)
A tree up to 60 ft. tall with bright green, silky twigs. Flower white, showy, fragrant, 2-3 in. across. Fruits dry, reddish, in a 1-2 in. long cone. It grows in wet, sandy soil in Oklahoma, Arkansas, Louisiana and eastern Texas. It is in the magnolia family *(Magnoliaceae)*.

Fig. 249

BLACK HUCKLEBERRY
(Gaylussacia baccata)
An erect, rigid shrub 1-4 ft. tall. The flowers are conic, ¼ in. long and pink in color. The fruit is a sweet, berry-like, shiny-black drupe. It grows in acid, swampy soil in Louisiana. It belongs to the heath family *(Ericaceae)*.

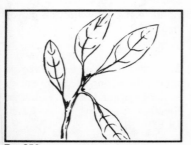

Fig. 250

CANDLEBERRY
(Myrica cerifera)
An evergreen shrub or small tree up to 35 ft. tall with waxy twigs. The flowers are inconspicuous. The fruit is globose, 1/8 in. in diameter, light green and covered with wax granules. It grows in low, swampy ground in Arkansas, Oklahoma, Louisiana and eastern Texas. It belongs to the wax-myrtle family *(Myricaceae)*.

248

1. Check both blade surfaces, if hairy or scaly on *both* surfaces, look for (KI 126), shown to the right, in the left-hand margin of page 251.

KI 125

2. If only the lower surface, or just its veins are hairy, find the symbol (KI 130), shown to the right, in the left-hand margin of page 259.

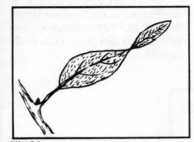

KI 126

KI 130

1. If the blade shape of your specimen is reverse tear-shape with smooth twigs and leaf stalks, it is DWARF WILLOW.

2. If the blade shape of your specimen is reverse tear-shape with sticky-hairy twigs and leaf-stalks, it is DWARF HUCKLE-BERRY.

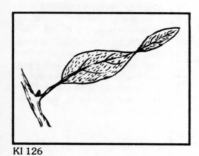

KI 126

3. If the blade is tear-, oval-, elliptic-, oblong- or triangle-shaped, find (KI 127), to the right, in the left-hand margin of page 253.

Fig. 251

DWARF WILLOW

(Salix tristis)

A shrub 1-3 ft. tall with numerous leafy branches. Inconspicuous flowers are borne in lax spikes. Fruit is a small capsule bearing cottony seeds. It grows on acid, sandy soil in Louisiana, Oklahoma and Arkansas. It belongs to the willow family *(Salicaceae)*.

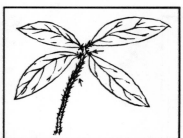

Fig. 252

DWARF HUCKLEBERRY

(Gaylussacia hirtella)

A hairy shrub 1-5 ft. tall. The urn-shaped flowers are ⅓ in. long and white to pink in color. The fruit is black, lustrous and hairy and about ⅓ in. in diameter. It grows in wet, acid soil in Louisiana and belongs to the heath family *(Ericaceae)*.

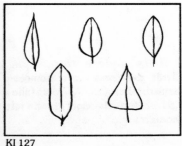

KI 127

252

1. If the leaves are silvery-scaly (note variable leaf shape in symbol to the right), it is BIG SALT-BUSH.

2. If the leaves are dense yellow and hairy (and the twigs as well), it is FRAGRANT CROTON.

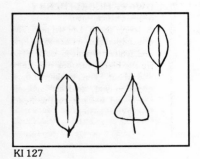

KI 127

3. If the leaves are non-scaly and the hairs non-yellow, follow (KI 128), at the right, in the left-hand margin of page 255.

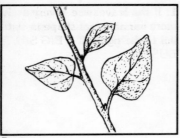
Fig. 253

BIG SALTBUSH
(Atriplex lentiformis)
A rounded shrub up to 5 ft. tall with many slender, flexulous branches. The inconspicuous tiny flowers are borne in crowded spikes. The tiny fruit is seed-like. It grows in alkaline desert soils in New Mexico. It belongs to the goosefoot family *(Chenepodiaceae)*.

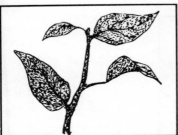
Fig. 254

FRAGRANT CROTON
(Croton suaveolens)
A rounded shrub about 3 ft. tall with densely hairy stems and fragrant foliage. The tiny flowers are inconspicuous, imperfect and without petals. The fruit is a rounded capsule about ¼ in. long. It grows on rocky bluffs and slopes in Trans-Pecos Texas near Fort Davis and also in Val Verde county. It belongs to the spurge family *(Euphorbiaceae)*.

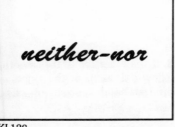

neither-nor

KI 128

254

1. If the older twigs are densely hairy but non-sticky to the touch, it is COLDENIA.

2. If the twigs are hairy and sticky, it is BEACH CROTON.

neither–nor

KI 128

3. If the older twigs are smooth, follow (KI 129), at the right, in the left-hand margin of page 257.

255

Fig. 255

COLDENIA
(Coldenia greggii)
A low, hairy, compact, rounded shrub from 4 in. to 2 ft. tall. The flowers are pink to magenta, and about ⅓ in. long and are borne in dense, terminal, rounded heads. The fruit is a dry, small nutlet. It grows on rocky hillsides and ravines in far western Texas and New Mexico. It belongs to the borage family *(Boraginaceae)*.

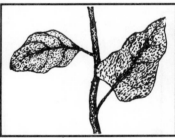

Fig. 256

BEACH CROTON
(Croton punctatus)
A semi-shrub with creeping woody stems and herbaceous stems up to 4 ft. tall. The inconspicuous, imperfect flowers are borne in few-flowered, densely hairy clusters. The fruit is a rounded, 3-lobed capsule about ⅓ in. in diameter. It grows in dune sand along the Louisiana and Texas coasts. It belongs to the spurge family *(Euphorbiaceae)*.

KI 129

256

1. If the leaves are elliptic, ¾-2 in. long, with plant height under 1 ft., it is DWARF BLUE-BERRY.

2. If the leaves are oblong (as shown to the right), 2-6 in. long with plant height 15-25 ft., it is WHITE WILLOW.

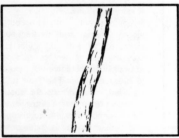

KI 129

3. If the leaves are oblong, 1-2 in. long with plant height 3-6 ft., it is TORREY CROTON.

4. If the leaves are tear-shaped to elliptic, 3-6 in. long with plant height to 20 ft. and growing in swamps, it is CORKWOOD.

Fig. 257

DWARF BLUEBERRY
(Vaccinium depressum)
A shrub less than 1 ft. tall with slender, finely hairy branches. The white, urn-shaped flowers are 1/5 in. long. The fruit is a red to purplish-black berry ⅓ in. in diameter. It grows in sandy pine forests in southern Louisiana. It belongs to the heath family *(Ericaceae)*.

Fig. 258

WHITE WILLOW
(Salix lasiolepis var. bracelinae)
An open-crowned small tree or shrub. Catkins lax, seeds cottony. It grows along streams in far western Texas and New Mexico. It is in the willow family *(Salicaceae)*.

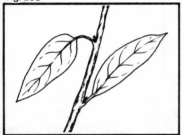

Fig. 259

TORREY CROTON
(Croton torreyanus)
A densely hairy shrub 3-6 ft. tall. Flower tiny, clusters dense, hairy 1½ in. long. Fruit a hairy capsule about ¼ in. in diameter. It grows on limestone soils in New Mexico and far western Texas. It is in the spurge family *(Euphorbiaceae)*.

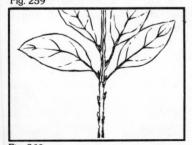

Fig. 260

CORKWOOD
(Leitneria floridana)
A shrub or small tree up to 20 ft. tall. Flowers inconspicuous, borne in elongate spikes 1-2 in. long. The fruit is olive-like (drupe), ¾ in. long, brown and leathery. It grows in swamps in southern Louisiana and southeastern Texas. It belongs to the corkwood family *(Leitneriaceae)*.

1. If the leaf-stalks of the *older* leaves are hairy, follow (KI 131), at the right, in the left-hand margin of pages 261 and 261A.

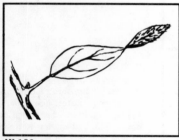

KI 130

2. If the leaf-stalks of the older leaves are hairless, find the symbol, (KI 133), to your right, in the left-hand margin of pages 265 and 265A.

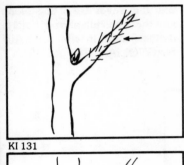

KI 131

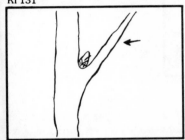

KI 133

1. If the leaves of your specimen are very large, 20 to 30 in. long, it is LARGE-LEAF MAGNOLIA.

2. If the leaves are 5-12 in. long and the leaf stalk hairs are not a definite color, it is MOUNTAIN MAGNOLIA.

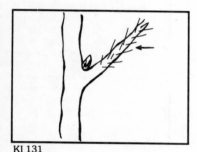

KI 131

3. If the longest leaves are more than 5 in. long (7 in.), the leaf-stalk hairs are reddish and the blade outline reverse egg-shape, it is SMALL PAW-PAW.

Fig. 261

LARGE-LEAF MAGNOLIA
(Magnolia macrophylla)
A broad-crowned tree up to 50 ft. tall. Flowers 10-18 in. across, white, fragrant with many short stamens. Fruits borne in a cone-like cluster 3 in. long. It grows along streams in eastern Arkansas and Louisiana. It is in the magnolia family *(Magnoliaceae)*.

Fig. 262

MOUNTAIN MAGNOLIA
(Magnolia pyramidata)
Slender tree up to 30 ft. tall. Flower white, 4 in. across, stamens many. Fruits borne in a rosy cluster 2½ in. long. It grows along streams in southeastern Louisiana and southeastern Texas. It is in the magnolia family *(Magnoliaceae)*.

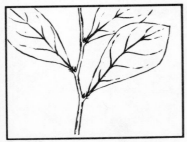

Fig. 263

SMALL PAW-PAW
(Asimina parviflora)
A shrub up to 12 ft. tall. Flowers purplish, 3/5 in. across. Fruit oval, about ¾-2 in. long. It grows along streams or in dry woods in southeastern Texas and southern Louisiana. It belongs to the custard-apple family *(Annonaceae)*.

4. If the longest leaves are more than 5 in. long (9 in.), the leaf stalk hairs are reddish and the blade outline oval to elliptic, it is SOUTHERN MAGNOLIA.

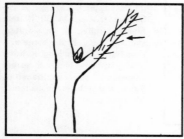

KI 131

5. If the leaves are all less than 5 in. long, go to (KI 132), at the right, in the left-hand margin of page 263.

Fig. 264

SOUTHERN MAGNOLIA

(Magnolia grandiflora)

Evergreen tree up to 90 ft. tall. Flowers cup-shaped, fragrant, white, 6-9 in. across. Rusty-haired fruits clustered into a 2-4 in. long cone. It grows in rich, moist soil in Texas, Arkansas, Louisiana and Oklahoma. It is in the magnolia family *(Magnoliaceae)*.

‹ 5″

KI 132

262A

1. If the leaves are tear-shaped, it is SILVER-LEAF OAK.

2. If the leaves are oval or some triangular, it is THIN-LEAF HACKBERRY.

KI 132

3. If the leaves are elliptic and the bark on the older twigs has shed in narrow flakes, it is MOUNTAIN LAUREL.

4. If the leaves are elliptic and the bark on the older twigs is smooth and tight, it is SHINGLE OAK.

263

Fig. 265

SILVER-LEAF OAK
(Quercus hypoleucoides)
A shrub or small tree up to 35 ft. tall. Flowers inconspicuous; fruit an acorn about 3/5 in. long. It grows in moist, montane canyons of Trans-Pecos Texas and New Mexico. It is in the beech family *(Fagaceae)*.

Fig. 266

THIN-LEAF HACKBERRY
(Celtis tenuifolia)
A shrub or small tree up to 24 ft. tall. Flowers inconspicuous, the fruit thin-fleshed, brown, orange or red, ⅓ in. in diameter. It grows in dry, rocky soils in Arkansas, Louisiana, eastern Oklahoma and eastern Texas. It is in the elm family *(Ulmaceae)*.

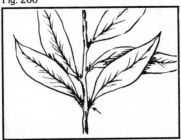

Fig. 267

MOUNTAIN LAUREL
(Kalmia latifolia)
See page 244.

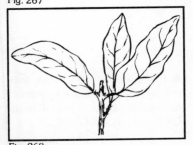

Fig. 268

SHINGLE OAK
(Quercus imbricata)
A tree up to 60 ft. tall. Flowers inconspicuous, the fruit of an acorn about 3/5 in. long. It grows along streams or on hillsides in Arkansas, eastern Oklahoma and northern Louisiana. It is in the beech family *(Fagaceae)*.

1. If the leaves are 1-2 in. long and reverse egg-shape, as pictured in the diagram to your right, it is TEXAS PERSIMMON.

2. If the leaves are tear-shaped, it is the CORTES CROTON.

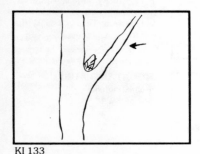

KI 133

3. If the leaves are reverse tear-shape and reverse egg-shape and up to 24 in. long, it is the UMBRELLA MAGNOLIA.

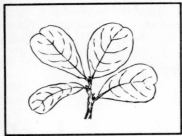

Fig. 269

TEXAS PERSIMMON
(Diospyros texana)
A gray, smooth-barked shrub or tree up to 40 ft. tall. Flowers urn-shaped, greenish-white, ⅓ in. long. Fruit pulpy, rounded, 1 in. in diameter. It grows on rocky hills in the western ⅔ of Texas. It belongs to the ebony family *(Ebenaceae)*

Fig. 270

CORTES CROTON
(Croton cortesianus)
A shrub 3-6 ft. tall with tiny flowers in dense, elongate spikes. The fruit is a rounded, hairy capsule, about ⅓ in. long. It is rare in the southern tip of Texas. It belongs to the spurge family *(Euphorbiaceae)*.

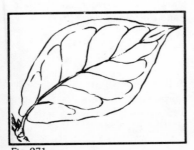

Fig. 271

UMBRELLA MAGNOLIA
(Magnolia tripetala)
Shrub or tree up to 40 ft. tall. Flowers cup-shaped, white, fragrant, 8-10 in. across. Fruits rose-colored in a cone. It grows in swamps or on mountain slopes in southeastern Oklahoma and central-southwestern Arkansas. It is in the magnolia family *(Magnoliaceae)*.

266

4. If the leaves are reverse tear-shape and 1-2 in. long, it is the DANGLEBERRY.

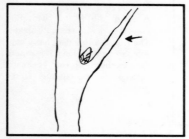

KI 133

5. If the leaves are oval or elliptic in outline, follow (KI 134), shown to the right, in the left-hand margin of page 267.

Fig. 272

DANGLEBERRY

(Gaylussacia frondosa)

A slender shrub 4-9 ft. tall. The few flowers are white to greenish-purple, 1/5 in. long and bell-shaped. The fruit is dark blue, fleshy and ⅓ in. in diameter. It grows in moist forests in Louisiana. It belongs to the heath family *(Ericaceae)*.

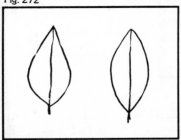

KI 134

1. If the leaves are all less than 1 in. long, it is the BLACK HUCK-LEBERRY.

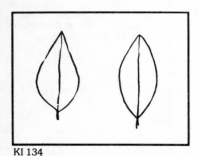

KI 134

2. If the longest leaves are more than 1 in. long, find drawing (KI 135), at the right, in the left-hand margin of page 269.

Fig. 273

BLACK HUCKLEBERRY
(Gaylussacia baccata)
See page 248.

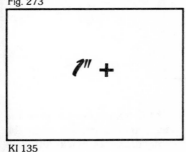

1" +

KI 135

1. If the leaves are hairy on the undersurface vein only and the blade is thin, it is WILLOW OAK.

2. If the leaves are hairy on the undersurface veins only and the blade is thick, it is THICK-LEAVED BLUEBERRY.

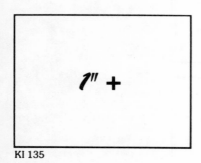

KI 135

3. If the leaves are hairy on the blade undersurface and the young branches black-dotted, it is STAGGER BUSH.

4. If the leaves are hairy on the blade undersurface and the branches are not dotted, find (KI 136), at the right, in the left-hand margin of page 271.

Fig. 274

WILLOW OAK
(Quercus phellos)
A tree up to 65 ft. tall. Male catkins lax, dense, hairy, yellowish-green. Fruit an acorn 3/5 in. long. It grows in bottomland forests in Arkansas, Louisiana, eastern Oklahoma and eastern Texas. It is in the beech family *(Fagaceae)*.

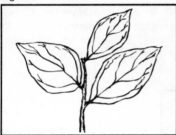

Fig. 275

THICKLEAVED BLUEBERRY
(Vaccinium fuscatum)
A shrub 3-9 ft. tall with pink to reddish, urn-shaped flowers 1/3 in. long. Fruit a dark berry about 2/5 in. in diameter. It grows in old fields and low pine forests in eastern Louisiana and Arkansas. It belongs to the heath family *(Ericaceae)*.

Fig. 276

STAGGER BUSH
(Lyonia mariana)
A slender shrub up to 6 ft. tall. Flowers cylindrical, nodding, white to pink and about 1/2 in. long. Fruit an oval-shaped capsule about 2/5 in. long. It grows in pine forests in Louisiana and eastern to south-central Texas. It is in the heath family *(Ericaceae)*.

no dots

1. If your specimen occurs in the high mountain forest below timberline in New Mexico and is 3-6 in. high, it is GRAY-LEAF WILLOW.

2. If your specimen is a tree growing on limestone hills in Texas, Oklahoma or Arkansas, it is SMOKE TREE.

no dots

KI 136

3. If your specimen is a tree or shrub growing in the rich bottomlands, sandy acid soils, or bogs of Texas, Arkansas, Louisiana or Oklahoma, go to (KI 137), shown to the right, in the left-hand margin of page 273.

Fig. 277

GRAY-LEAF WILLOW
(Salix glauca)
A shrub 3-6 ft. tall. The tiny, imperfect flowers are borne in lax spikes (catkins) which are hairy and 1½-2 in. long. The fruit is a tiny capsule bearing silky seeds. It grows in boggy meadows in the upper mountain reaches of northern New Mexico. It belongs to the willow family *(Salicaceae)*.

Fig. 278

SMOKE TREE
(Cotinus obovatus)
A straggly shrub or small tree up to 30 ft. tall with yellow wood and strong-scented juice. Flower greenish-yellow ⅛ in. across. Fruit flat, kidney-shaped, ¼ in. long. It grows on limestone hills in Texas, Oklahoma and Arkansas. It is in the sumac family *(Anacardiaceae)*.

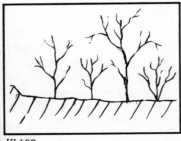
KI 137

272

1. If the leaves are 1 in. or less wide, it is a HUCKLEBERRY.

2. If the broadest leaves are 3 in. wide, and the leaf tip is blunt, as shown in the diagram to the right, it is BLACK-GUM.

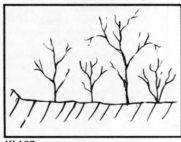

KI 137

3. If the broadest leaves are 3 in. wide and the leaf tip is sharp, as shown in the diagram to the right, it is PERSIMMON.

Fig. 279

HUCKLEBERRY
(Gaylussacia and Vaccinium spp.)
A vegetatively-similar species group of shrubs with small, white to pink, urn-shaped flowers and blue, fleshy fruits. They occur in acid soils in the eastern portion of our 5-state area. They are in the heath family *(Ericaceae)*.

Fig. 280

BLACK-GUM
(Nyssa sylvatica)
A large tree up to 90 ft. tall with horizontally spreading branches. The small greenish flowers are imperfect and not too distinctive. The fruit is olive-like, bluish-black and ½ in. long. It grows in moist, rich soils in eastern Texas, Oklahoma, Arkansas and Louisiana. It belongs to the dogwood family *(Cornaceae)*.

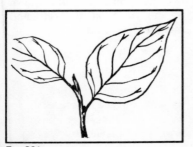

Fig. 281

PERSIMMON
(Diospyros virginiana)
A thicket-forming tree up to 30 ft. tall. Flower greenish-yellow, petals recurved, calyx fleshy, about ½ in. across. Fruit fleshy, orangish, pulpy, calyx persistent, about ¾-1½ in. in diameter. It grows in old fields or forest clearings in the eastern half of Texas and Oklahoma and in Arkansas. It is in the ebony family *(Ebenaceae)*.

1. Check for stalks on the leaves of your specimen. If most are absent (sessile leaves), go to (KI 139), at the right, in the left-hand margin of page 277.

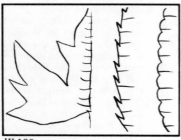

KI 138

2. If the leaf stalks are apparent on most leaves, even if short, find the symbol (KI 142), shown to the right, in the left-hand margin of page 283.

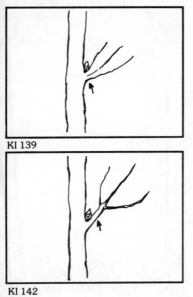

KI 139

KI 142

1. If the stem has regularly spaced ring nodes, as shown to the right, and the stem segments between these nodes are channeled on one side, it is CANE BAMBOO.

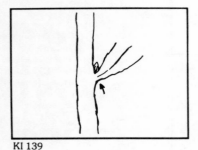

KI 139

2. If regular ring nodes are absent, proceed to (KI 140), at the right, in the left-hand margin of pages 279 and 279A.

Fig. #282

CANE BAMBOO
(Arundinaria gigantea)
A woody grass of the bamboo tribe, sending many stems from underground stems up to 30 ft. tall to form dense colonies called cane brakes. The grass flowers and seeds seldom form and then inconspicuous and group into scaley "spikelets." Stems ringed at the nodes and hollow, forming fishing poles when stripped of their branchlets. Plants of swampy ground in eastern Oklahoma, Arkansas, Louisiana and eastern Texas or in sandy soil as an escape from cultivation. It belongs to the grass family *(Graminae)*.

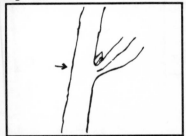

KI 140

1. If the leaf outline is fan-shaped and lobed, as shown to the right, with its upper surface gland-dotted, it is QUININE BUSH.

2. If the leaf is similar to that of QUININE BUSH above but lacks upper surface glands, it is APACHE PLUME.

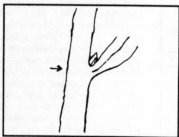

KI 140

3. If the leaf outline is reverse tear-shape and the margin toothed, as shown to the right, it is DAMIANA.

Fig. 283

QUININE BUSH
(Cowania stansburiana)
Spreading, evergreen shrub 1-6 ft. tall. Flower resinous, strong-scented, white, ¾ in. across. Fruit a long, slender plume. It grows on open slopes in the foothills and mountains of New Mexico. It is in the rose family *(Rosaceae)*.

Fig. 284

APACHE PLUME
(Fallugia paradoxa)
A dense, slender shrub up to 6 ft. tall. The flowers are showy, white, 1 in. across, bearing many stamens. The fruits are long plumes, crowded into reddish clusters. It grows in dry, rocky soil in the mountains of New Mexico eastward into central and northwestern Texas. It belongs to the rose family *(Rosaceae)*.

Fig. 285

DAMIANA
(Turnera diffusa var. aphrodisiaca)
An aromatic spreading shrub mostly 2 ft. tall, occasionally up to 6 ft. in height. The flower is small, yellow and ⅓ in. long. The fruit is a rounded, dry capsule, ¼ in. in diameter. It grows in brush on hillsides along the Rio Grande River in southern Texas. It belongs to the turnera family *(Turneraceae)*.

4. If the leaves are narrow, few-toothed and the margins parallel (also look for white, hairy surfaces), it is COYOTE WILLOW.

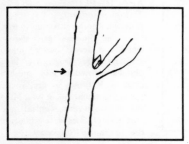

KI 140

5. If the leaf shape is elliptical, oval or reverse-oval, as illustrated in (KI 141), to the right, in the left-hand margin of page 281.

Fig. 286

COYOTE WILLOW
(Salix argophylla)
A thicket-forming shrub or small tree 3-18 ft. tall. Catkins ¾-1¾ in. long. It grows along streams and wet ditches in New Mexico and far western Texas. It is in the willow family *(Salicaceae).*

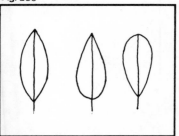

KI 141

1. If the leaves are heavily hairy and the margin few-toothed, it is JOHNSTON'S BERNARDIA.

2. If the leaves are heavily hairy and the margin regularly toothed, it is the SOUTH-WESTERN BERNARDIA.

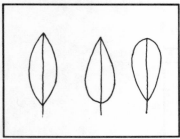

KI 141

3. If the leaves are sparsely hairy on the undersurface (mostly on the veins) and the margin is sawtoothed, it is a BLUE-BERRY.

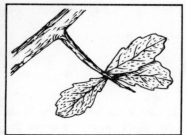

Fig. 287

JOHNSTON'S BERNARDIA
(Bernardia obovata)
Shrub up to 3 ft. tall. Flower tiny, hairy. Fruit a capsule ⅓ in. long. It grows on stony soils in far western Texas. It belongs to the spurge family *(Euphorbiaceae).*

Fig. 288

SOUTHWESTERN BERNARDIA
(Bernardia myricaefolia)
A dense shrub up to 7 ft. tall. Flower tiny, hairy. Fruit a 3-lobed capsule about ½ in. in diameter. It is widespread on rocky sites in central and southern Texas. It is in the spurge family *(Euphorbiaceae).*

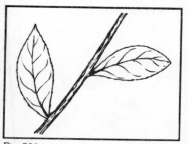

Fig. 289

BLUEBERRY
(Vaccinium amoenum and *virgatum)*
Vegetatively similar species, the former 3-6 ft. tall and the latter up to 3 ft. tall. The flowers are urn-shaped and pink-tinged to pink in color, ranging from ¼-½ in. in length. The fruit is a berry, black in color and 2/5 in. in diameter for both species. Both grow in acid soil in Arkansas, Louisiana and eastern Texas and Oklahoma. They belong to the heath family *(Ericaceae).*

1. Check the base of the blade (the point where it joins the stalk). If the two blade halves do not meet on the stalk or are uneven, as shown in (KI 143), to the right, in the left-hand margin of page 285.

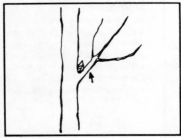

KI 142

2. If the blade halves meet at the stalk, find the symbol (KI 146), to the right, in the left-hand margin of page 291.

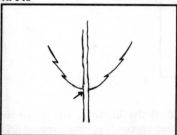

KI 143

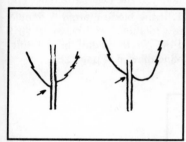

KI 146

1. If the blade margin is double saw-toothed, as shown in (KI 144), to the right, in the left-hand margin of page 287.

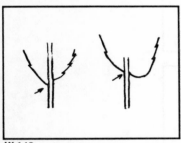

KI 143

2. If the blade margin is single saw-toothed or scalloped, find the symbol (KI 145), at the right, in the left-hand margin of pages 289 and 289A.

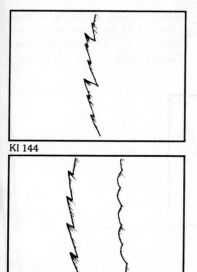

KI 144

KI 145

1. If the leaf outline is reverse egg-shape, it is the SEPTEMBER ELM.

2. If the leaf outline is oval to elliptic and the older leaves are short-hairy on their undersurfaces, it is SLIPPERY ELM.

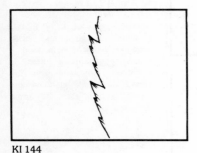

KI 144

3. If the leaf outline is oval to elliptic and the older leaves are hairless on their undersurfaces, it is AMERICAN ELM.

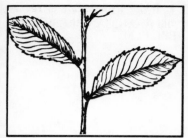

Fig. 290

SEPTEMBER ELM
(Ulmus serotina)
A tree up to 60 ft. tall. Flower inconspicuous. Fruit seed-like, winged, ½ in. long. It grows on limestone hills-riverbanks in eastern Oklahoma and northwestern Arkansas. It is in the elm family *(Ulmaceae)*.

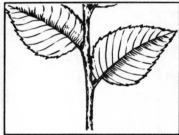

Fig. 291

SLIPPERY ELM
(Ulmus rubra)
Spreading tree up to 60 ft. tall. Flowers inconspicuous; fruits seed-like, winged, ½ in. long. It grows in streamside forests in Arkansas, Louisiana and the eastern third of Oklahoma and Texas. It is in the elm family *(Ulmaceae)*.

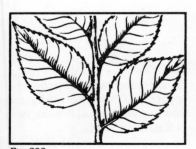

Fig. 292

AMERICAN ELM
(Ulmus americana)
A mostly 60 ft. tall tree with gradually spreading branches to form a vase shape. Flowers inconspicuous. Fruits seed-like, winged and ½ in. long. It grows in streamside forests in Arkansas, Louisiana and the eastern third of Texas and Oklahoma. It belongs to the elm family *(Ulmaceae)*.

288

1. If the leaves are hairless, it is the AMERICAN BASSWOOD.

2. If the leaves are hairy only in the vein angles on the undersurface of the leaf, it is the FLORIDA LINDEN.

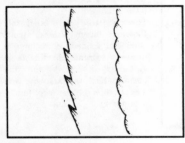

KI 145

3. If the leaves are hairy on their undersurfaces and the margin is saw-toothed and you have found your specimen growing in northern Arkansas, it is the BEE TREE.

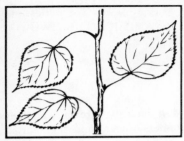

AMERICAN BASSWOOD
(Tilia americana)
Tree up to 120 ft. tall. Flowers yellowish-white, small, attached to center of a bract 2-5 in. long. Fruit brown-hairy, ⅓ in. in diameter. It grows in rich soil along streams in eastern Oklahoma, northeast Texas and Arkansas. It belongs to the linden family *(Tiliaceae)*.

Fig. 293

FLORIDA LINDEN
(Tilia floridana)
Tree up to 60 ft. tall. Flowers, fruits as above except bract 3-6 in. long. It grows in Louisiana, Arkansas, central to eastern Texas and eastern Oklahoma on rich, forested soils. It belongs to the linden family *(Tiliaceae)*.

Fig. 294

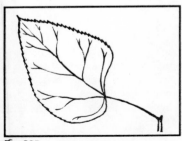

BEE TREE
(Tilia heterophylla)
A tree up to 80 ft. tall. Flower, fruit, bract characters same as for the lindens above. It occurs in northern Arkansas. It is in the linden family *(Tiliaceae)*.

Fig. 295

4. If the leaves are hairy on their undersurfaces, and the margin saw-toothed, and you have found your specimen in Texas, Louisiana or southwestern Arkansas, it is the CAROLINA BASSWOOD.

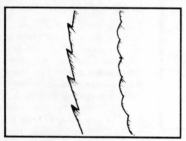

KI 145

5. If the leaves are hairy on their undersurfaces and the margin is scalloped, it is WITCH HAZEL.

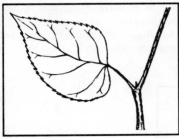

Fig. 296

Fig. 297

CAROLINA BASSWOOD
(Tilia caroliniana)
A tree up to 95 ft. tall. Floral, fruit and bract characters same as for above lindens. It grows in lowland forests in western Louisiana, southwestern Arkansas and central to eastern Texas. It belongs to the linden family *(Tiliaceae)*.

WITCH HAZEL
(Hamamelis virginiana)
Tall shrub or small tree up to 30 ft. tall. Flower yellow, petals long, narrow twisted. Fruit a 2-beaked, woody capsule, ½ in. long. It grows along streams in Arkansas, Louisiana and Oklahoma. East Texas form has red-tinged petals. It is in the witch hazel family *(Hamamelidaceae)*.

1. If the older leaves are hairless on both surfaces (check under-surface veins also), find (KI 147), to the right and in the left margin of page 293.

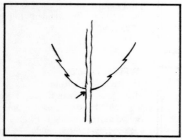

KI 146

2. If the older leaves are hairy on either surface (or both) or on the undersurface veins only, find the symbol (KI 170), at the right, in the left-hand margin of page 339.

non-hairy

KI 147

hairy

KI 170

1. Check the leaf stalk of your specimen and determine if it is flattened or twisted or roundish in cross-section. If the stalk is flat or twisted proceed to (KI 148), on page 295.

non-hairy

KI 147

2. If the leaf stalk is round, find the symbol (KI 149), to the right, in the left-hand margin of page 297.

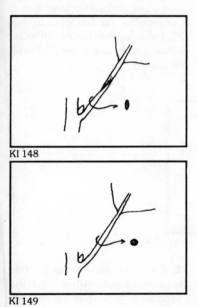

KI 148

KI 149

1. If the leaves are triangular in outline, it is the COTTON-WOOD.

2. If the leaves are broad tear-shaped, it is the PEACH-LEAF WILLOW.

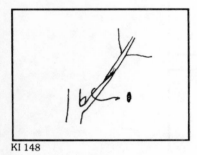

KI 148

3. If the leaves are egg-shaped or elliptical in outline and the bark is white, it is QUAKING ASPEN.

4. If the leaves are egg-shaped to elliptical in outline and the bark is gray, it is OKLAHOMA ALDER.

295

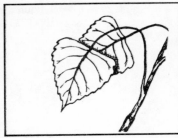

Fig. 298

COTTONWOOD
(Populus spp.)

Mostly large trees that are often separated into species by range. The tiny, imperfect flowers are borne in lax spikes (catkins). The fruit is a small, rounded capsule bearing cottony seeds. They grow along streams throughout our 5-state area. They belong to the willow family *(Salicaceae)*.

Fig. 299

PEACH-LEAF WILLOW
(Salix amygdaloides)

Tree up to 40 ft. tall. Flowers inconspicuous in yellowish-green catkins 1-3 in. long. Fruit a tiny capsule. It grows around lakes or along streams in the lower mountains of far western Texas into New Mexico. It is in the willow family *(Salicaceae)*.

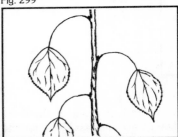

Fig. 300

QUAKING ASPEN
(Populus tremuloides)

White-barked tree up to 40 ft. in colonies from root sprouts. Nonflowering in our area. It grows in the high mountains of New Mexico and far western Texas. It is in the willow family *(Salicaceae)*.

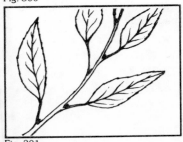

Fig. 301

OKLAHOMA ALDER
(Alnus maritima)

Thicket-forming shrub up to 15 ft. tall or small tree. Flowers imperfect, tiny, the male catkin 2 in. long, the female "cone" 1/6 in. long. It grows in wet soil in southern Oklahoma. It is in the birch family *(Betulaceae)*.

1. Observe the leaf stalks of the *older* leaves for hairness, if hairy, go to (KI 150), at the right, in the left-hand side of page 299.

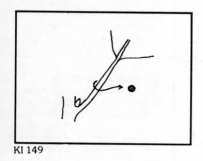

KI 149

2. If the leaf stalks of the *older* leaves are hairless, look for the symbol (KI 153), shown to the right, in the left-hand margin of page 305.

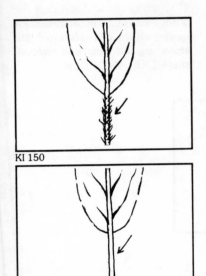

KI 150

KI 153

1. If the undersurface of the leaf is white and the stalks are winged in the manner shown in the drawing to the right, it is CAROLINA WILLOW.

2. If the leaf undersurface is green, the stalks unwinged but the margin double saw-toothed, it is MEXICAN ALDER.

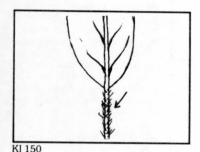

KI 150

3. If the undersurface of the leaf is green, the stalks wingless and the margin single saw-toothed or scalloped, go to (KI 151), at the right, in the left-hand margin of page 301.

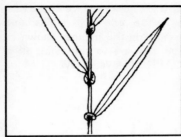

Fig. 302

CAROLINA WILLOW

(Salix caroliniana)

A shrub or small tree up to 25 ft. tall. The tiny imperfect flowers are borne in lax spikes (catkins) which are up to 4 in. long and yellowish. The fruit is a small capsule bearing cottony seeds. It grows around ponds or along streams in Louisiana, eastern Oklahoma, western Arkansas and south-central Texas. It belongs to the willow family *(Salicaceae)*.

Fig. 303

MEXICAN ALDER

(Alnus oblongifolia)

Shrub or tree up to 30 ft. tall. Flowers imperfect, the male catkin yellow-orange, 2 in. long, the female "cone" 1 in. long. It grows along streams in low altitudes in the western mountains of New Mexico. It is in the birch family *(Betulaceae)*.

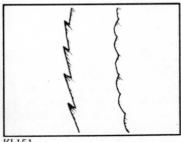

KI 15J

1. If the marginal teeth are blunted, thus the margin appearing scalloped, it is GEORGIA SWAMP HOLLY.

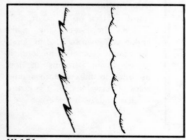

KI 151

2. If the teeth are sharp, go to (KI 152), at the right, in the left-hand margin of page 303.

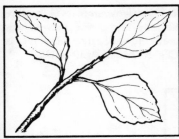

Fig. 304

GEORGIA SWAMP HOLLY
(Ilex longipes)
Shrub or rarely a tree up to 25 ft. tall. Flower, small, imperfect, with petals 4, greenish-white; fruit red, cherry-like, ⅓ in. in diameter. It grows in low, sandy soil in southern Louisiana and southeastern Texas. It is in the holly family *(Aquifoliaceae)*.

KI 152

302

1. If some leaves are elliptic and some have parallel margins, as shown to the right, it is a BUCKTHORN.

2. If the leaves are oval or elliptic only and some leaves lack teeth toward the base, it is COASTAL LEUCOTHOE.

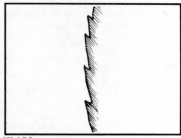
KI 152

3. If the leaves are oval-elliptic and the teeth very small, it is SWEETSPIRE.

4. If the leaves are oval-elliptic and the teeth are quite evident, it is WATER ELM.

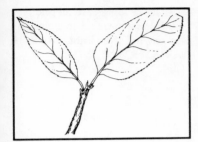

Fig. 305

BUCKTHORN

(Rhamnus spp.)

Shrubs or small trees with greenish-yellow, small flowers and red to black, cherry-like fruits around ⅓ in. in diameter. They are found in seepage areas or along streams in all states of our area. They belong to the buckthorn family *(Rhamnaceae)*.

Fig. 306

COASTAL LEUCOTHOE

(Leucothoe axillaris)

An evergreen shrub up to 6 ft. tall with recurved branches. The flower clusters are dense with white, cylindric flowers ¼ in. long. The fruit is a roundish capsule ⅓ in. long. It grows in acid swamps near the coast in southeastern Louisiana. It belongs to the heath family *(Ericaceae)*.

Fig. 307

SWEETSPIRE

(Itea virginica)

Shrub about 7 ft. tall. Flower white, fragrant, ½ in. across, in terminal clusters 2-5 in. long. Fruit a bi-groved capsule 1/6 in. long. It grows in seeps or along streams in Louisiana and eastern Texas. It is a member of the saxifrage family *(Saxifragaceae)*.

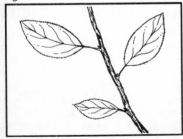

Fig. 308

WATER ELM

(Planera aquatica)

Shrub or small tree up to 30 ft. tall. Flower inconspicuous; fruit is leathery, warty, about ⅓ in. long. It grows in swamps, seeps or wet lowlands in Arkansas, Louisiana and eastern Oklahoma and Texas. It is in the elm family *(Ulmaceae)*.

304

1. If the leaves are lobed, the lobes smooth or saw-toothed, go to (KI 154), at the right, in the left-hand margin of page 307.

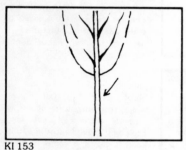

KI 153

2. If the leaves are unlobed but scalloped or toothed, find the picture (KI 157), to the right, in the left-hand margin of page 313.

KI 154

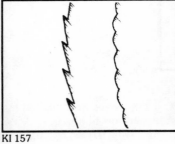

KI 157

1. If the lobe tips are blunt, 7-11 in number and the leaf 5-9 in. long, it is WHITE OAK.

2. If the lobe tips are blunt, 4 in number and the leaf 2-5 in. long, it is LACEY OAK.

KI 154

3. If the lobe tips are sharp, as pictured in (KI 155) to the right, go to this picture in the left-hand margin of page 309.

Fig. 309

WHITE OAK
(Quercus alba)
Large tree up to 150 ft. tall. Flowers inconspicuous, imperfect, the male catkin yellowish, hairy, 3 in. long; fruit an acorn 1 in. long. It grows on rich soil in Arkansas, Louisiana, eastern Texas and Oklahoma. It is in the beech family *(Fagaceae)*.

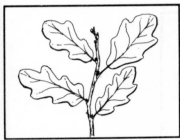

Fig. 310

LACEY OAK
(Quercus glaucoides)
Shrubs or small trees up to 40 ft. tall with gray-green foliage. Flowers imperfect, the male catkin loose, 2 in. long; fruit an acorn ¾ in. long. It grows on limestone outcrops in central Texas. It is in the beech family *(Fagaceae)*.

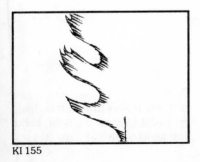

KI 155

1. If the leaves are ¼-2 in. long and some of the teeth are blunt, it is COMMON NINEBARK.

2. If the leaves are ½-2 in. long and all of the teeth are sharp, it is MOUNTAIN NINEBARK.

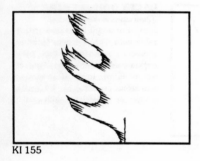

KI 155

3. If the leaves are 2-10 in. long, go to (KI 156), at the right, in the left-hand margin of page 311.

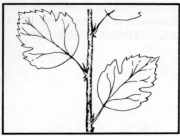

Fig. 311

COMMON NINEBARK
(Physocarpus opulifolius)
Shrub 3-9 ft. tall with shreddy bark and recurved branches. Flower white to pink 2/5 in. across in dense clusters 2 in. wide. Fruit a dry follicle ½ in. long. It grows on sandy or rocky soil in Arkansas. It is in the rose family *(Rosaceae)*.

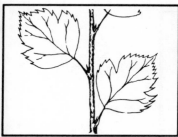

Fig. 312

MOUNTAIN NINEBARK
(Physocarpus monogynus)
Shrub up to 3 ft. tall, bark shreddy, brown. Flowers and fruits as above, except follicle ¼ in. long. It grows on open, rocky slopes in the mountains of New Mexico and far western Texas. It belongs to the rose family *(Rosaceae)*.

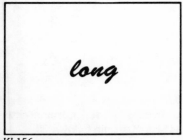

long

KI 156

310

1. Count the lobes of several leaves. If they range from 3-11, it is RED OAK.

2. If the lobes are only 7 in number, it is BLACK OAK.

long

KI 156

3. If the lobes range from 7-9, it is SHUMARD'S OAK.

Fig. 313

RED OAK

(Quercus rubra or *falcata* and hybrids)*

Trees up to 150 ft. tall. Male catkin 3-5 in. long; fruit an acorn ½-1 in. long. They grow on rich soil in eastern Texas and Oklahoma and Arkansas and Louisiana. They belong to the beech family *(Fagaceae).*

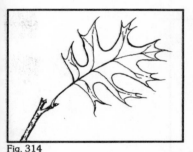

Fig. 314

BLACK OAK

(Quercus velutina)

Tree up to 50 ft. tall with coarse, black bark. Male catkin, hairy, 3-6 in. long. Fruit an acorn 1 in. long. It grows in upland forests in Louisiana, Arkansas, eastern Oklahoma and Texas in poor soil. It belongs to the beech family *(Fagaceae).*

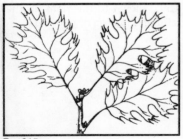

Fig. 315

SHUMARD'S OAK

(Quercus shumardii)

Tree up to 60 ft. tall. Male catkin 6-7 in. long. Fruit an acorn about 1 in. long. It grows along streams or damp hillsides in Oklahoma, Arkansas, eastern and central Texas. It belongs to the beech family *(Fagaceae).*

312

1. If the leaves are reverse tear-shape or reverse egg-shape, as shown in (KI 158), to the right, find this drawing in the left-hand margin of page 315.

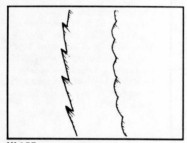

KI 157

2. If the leaves are egg-shape, elliptical, square, triangular or tear-shape, find this symbol (KI 159), shown to the right, in the left-hand margin of page 317.

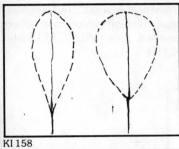

KI 158

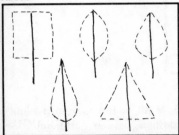

KI 159

1. If the leaves are toothed toward the apex and smell like shaving lotion when crushed, it is WAX-MYRTLE.

2. If the leaves are toothed toward the apex, but do not smell like shaving lotion when crushed, it is SWEET PEPPER-BUSH.

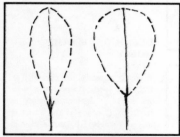

KI 158

3. If all of the margin is blunt-toothed, it is DECIDUOUS SWAMP HOLLY.

4. If all of the margin is finely sharp-toothed, it is SNAKE-WOOD.

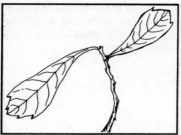
Fig. 316

WAX-MYRTLE
(Myrica spp.)
Evergreen shrub or sometimes a tree up to 40 ft. tall. Flowers imperfect, borne in catkins ¼-¾ in. long. The fruit is white, waxy and ⅛ in. in diameter. It grows in wet meadows, swamps and forests in eastern Oklahoma, Texas, Arkansas and Louisiana and belongs to the bayberry family *(Myricaceae)*.

SWEET PEPPER-BUSH
(Clethra alnifolia)
See page 404.

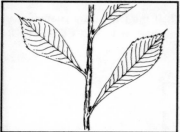
Fig. 317

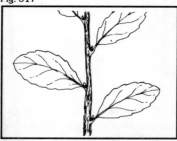
Fig. 318

DECIDUOUS SWAMP HOLLY
(Ilex decidua)
A shrub or small tree up to 30 ft. tall. The white flowers are ¼ in. wide. The distinctive fruit is rounded, fleshy, red and about ¼ in. in diameter. It grows in rich soil in ravines, stream-side forests and swamps in Arkansas and Louisiana and the eastern half of Oklahoma and Texas. It belongs to the holly family *(Aquifoliaceae)*.

SNAKEWOOD
(Colubrina texensis)
A thicket-forming shrub 3-6 ft. tall with twisting, snake-like branches. Flowers greenish-yellow, about ⅓ in. across. Fruit olive-like, dark brown and ⅓ in. in diameter. It grows in New Mexico and the western half of Texas. It belongs to the buckthorn family *(Rhamnaceae)*.

Fig. 319

316

1. If some of the leaves are curved in the manner illustrated to the right, it is HACKBERRY.

2. If the leaves are triangular in outline, it is BLOOD-OF-CHRIST.

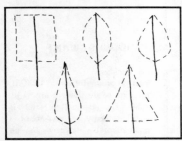

KI 159

3. If the leaves are square, as shown to the right, it is the TULIP TREE.

4. If the leaves are egg-shape, elliptical or tear-shaped, go to (KI 160), at the right, in the left-hand margin of page 319.

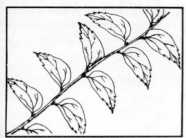

Fig. 320

HACKBERRY
(Celtis occidentalis)
A shrub to large trees up to 100 ft. tall with corky ridges or warts on the younger stems. The inconspicuous, small flower is greenish. The fruit is a thin-fleshed cherry (drupe) that is reddish-orange to black and ¼ in. in diameter. It grows in various habitats but mostly along streams in Oklahoma, Arkansas, Louisiana and the panhandle of Texas. It belongs to the elm family *(Ulmaceae)*.

Fig. 321

BLOOD-OF-CHRIST
(Jatropha cardiophylla)
A weak, low shrub up to 3 ft. tall. The cylindric flower is red and about ¼ in. long. The fruit is greenish-brown, rounded, ridged and ½ in. in diameter. It grows on dry, sandy-rocky soil in southern New Mexico. It belongs to the spurge family *(Euphorbiaceae)*.

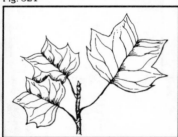

Fig. 322

TULIP TREE
Liriodendron tulipifera)
A long-trunked tree up to 120 ft. tall. Flowers handsome, yellowish-green, 3-4 in. across and many stamened. Fruits borne in cone-like clusters about 3 in. long. It grows on rich forest soil in Louisiana and Arkansas. It belongs to the magnolia family *(Magnoliaceae)*.

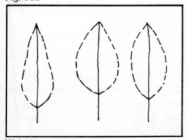

KI 160

1. If all of the leaves are narrowly tear-shaped and the surfaces are sticky to the touch, it is STICKY BACCHARIS.

2. If all of the leaves are narrowly tear-shaped and the surfaces are not sticky, it is the NARROW-LEAVED COTTONWOOD.

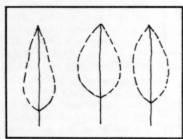

KI 160

3. If the leaves are egg-shape, elliptic or some (but not all) are tear-shaped, look for the drawing to the right (KI 161), in the left-hand margin of page 321.

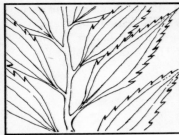

Fig. 323

STICKY BACCHARIS

(Baccharis glutinosa)

Slender-stemmed shrub up to 10 ft. tall with angled-striate branchlets. Flowers yellow, aster-like, in small heads 1/5 in. long. Fruit seed-like. It grows in colonies along streams in New Mexico and western Texas. It is in the aster family *(Compositae)*.

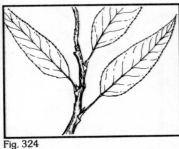

Fig. 324

NARROW-LEAVED COTTONWOOD

(Populus angustifolia)

A tree up to 60 ft. tall with a narrow crown. Flowers imperfect, borne in brownish catkins, 1-4 in. long. Fruit a small capsule bearing cottony seeds. It grows along mountain streams in New Mexico and far western Texas and belongs to the willow family *(Salicaceae)*.

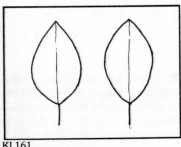

KI 161

320

1. If the teeth are spiny, it is AMERICAN HOLLY.

2. If the teeth are tiny saw-toothed and the young branches are red, smooth-barked, it is the TEXAS MADRONE.

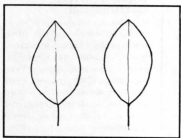

KI 161

3. If the teeth are neither spiny nor the branches red-barked, proceed to drawing (KI 162), shown to the right, in the left-hand margin of page 323.

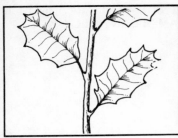

Fig. 325

AMERICAN HOLLY
(Ilex opaca)
Usually a small tree up to 50 ft. tall with evergreen foliage and stout branches. Flowers small, white or yellowish. Fruit prominent, red, fleshy, oval-shaped and ⅓-½ in. long. It grows along streams or in moist forests in eastern Oklahoma and Texas and Arkansas and Louisiana. It belongs to the holly family *(Aquifoliaceae)*.

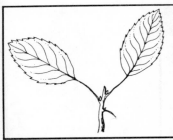

Fig. 326

TEXAS MADRONE
(Arbutus xalapensis)
Evergreen, crooked-branched, thicket-forming tree up to 30 ft. tall with pinkish, smooth bark that peels in papery sheets. Flowers urn-shaped, white to pinkish and ¼ in. long. Fruit fleshy, waxy, dark red to yellow and about ⅓ in. in diameter. It grows on wooded, rocky hills in southeastern New Mexico and central to western Texas. It is in the heath family *(Ericaceae)*.

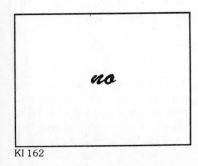

no

KI 162

322

1. If some or all of the teeth are blunt-tipped, look for drawing (KI 163), at the right, in the left-hand margin of page 325.

KI 162

2. If all of the teeth are sharp-tipped, find the illustration (KI 164), in the right, in the left-hand margin of page 327.

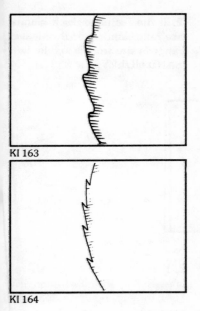

KI 163

KI 164

1. If some of the tooth tips are blunt, some sharp, and end in tiny glands, it is NEW JERSEY TEA.

2. If the tooth tips lack glands, are all blunt and the leaves range in size from 1 to 3 in., it is SAND HOLLY.

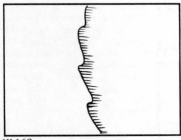

KI 162

3. If the tooth tips lack glands, are all blunt and the leaves are 1 in. or less long, it is EVERGREEN SWAMP HOLLY.

Fig. 327

NEW JERSEY TEA
(Ceanothus herbaceus)
Slender shrub up to 3 ft. tall. Flowers small, white in showy flat-topped clusters. Fruit a dry, 3-lobed capsule 2/5 in. long. It grows in open brushy areas in Arkansas, Louisiana, Oklahoma and west to central to the panhandle regions of Texas. It is in the buckthorn family *(Rhamnaceae)*.

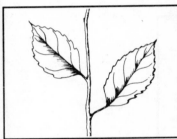

Fig. 328

SAND HOLLY
(Ilex ambigua)
A shrub or rarely a tree up to 18 ft. tall. The flower is tiny, white and hairy. The fruit is cherry-like, dark red and ¼ in. in diameter. It grows in sandy, streamside forests in Arkansas, Louisiana and eastern Oklahoma and Texas. It belongs to the holly family *(Aquifoliaceae)*.

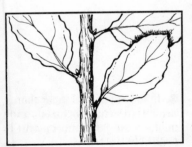

Fig. 329

EVERGREEN SWAMP HOLLY
(Ilex vomitoria)
A dense, evergreen, thicket-forming shrub or small tree up to 25 ft. tall. The flower is tiny and white. The fruit is cherry-like, red and ¼ in. in diameter. It grows in moist forests in south-central and southeastern Texas, Louisiana and Arkansas. It belongs to the holly family *(Aquifoliaceae)*.

1. If the leaves are 1 in. or less long and as broad as they are long, as shown to the right, it is the UTAH SERVICE-BERRY.

2. If the leaves are 1 in. or less long and longer than they are broad, as shown to the right, it is WHORTLE-BERRY.

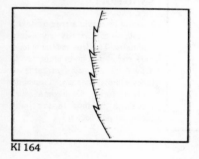

KI 164

3. If the leaves are longer than 1 in., find (KI 165), pictured to the right, in the left-hand margin of page 329.

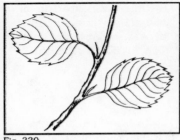

Fig. 330

UTAH SERVICE-BERRY
(Amelanchier utahensis)
A bushy shrub or a small tree up to 25 ft. tall. Flower white, more than 1 in. across and bearing many stamens. The fruit is a little apple (pome) that is blackish and ⅓ in. in diameter. It grows on steep limestone slopes in the far western mountains of Texas and westward into New Mexico. It belongs to the rose family *(Rosaceae)*.

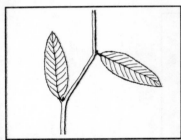

Fig. 331

WHORTLE-BERRY
(Vaccinium myrtillus)
Very slender-branched, open shrub up to 2 ft. tall. Flower urn-shaped, pinkish and ¼ in. long. The fruit is a blue berry ⅓ in. in diameter. It grows in the spruce zone in the mountains of northern New Mexico in our area. It belongs to the heath family *(Ericaceae)*.

KI 165

328

1. If the young twigs (at the tips of the branches) are hairy, as shown in (KI 166), to the right, proceed to this picture in the left-hand margin of page 331.

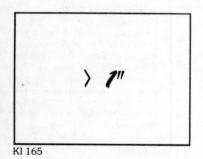

KI 165

2. If the young twigs are hairless, find the drawing (KI 168), shown to the right, in the left-hand margin of page 335.

KI 166

KI 168

1. If your specimen grows along streams only in Northwestern New Mexico, it is the ALDER-LEAVED SERVICE-BERRY.

KI 166

2. If your specimen grows in Texas, Oklahoma, Arkansas or Louisiana, find illustration (KI 167), shown to the right, in the left-hand margin of page 333.

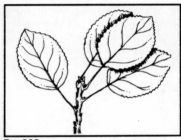
Fig. 332

ALDER-LEAVED SERVICE-BERRY

(Amelanchier alnifolia)

A thicket-forming, prostrate shrub, dwarfed shrub or a tree up to 6 ft. tall. Flower showy, fragrant, white, many stamened, 1-½ in. across. Fruit fleshy, purple to black, 3/5 in. in diameter. It glows along canyon streams in northwestern New Mexico. it belongs to the rose family *(Rosaceae).*

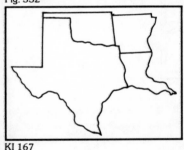
KI 167

1. If the leaves of your specimen range in length from 4 to 6 in., it is the HORTULAN PLUM.

2. If the leaves range in length from 1 to 5 in. and the bark is gray, smooth and thin, it is the SMOOTH ALDER.

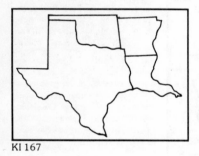

KI 167

3. If the leaves range in length from 2 to 5 in. and the bark is gray to black and pressured into ridges with age, it is the SHADBLOW.

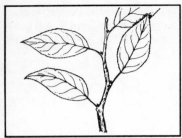

Fig. 333

HORTULAN PLUM
(Prunus hortulana)
A shrub or tree up to 30 ft. tall. The white flower is showy, up to 1 in. across and many-stamened. The fruit is a red to yellow-skinned plum and 1 in. long. It is rare as a species in the wild in Louisiana and Arkansas. It belongs to the rose family *(Rosaceae)*.

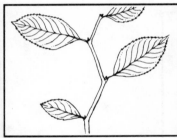

Fig. 334

SMOOTH ALDER
(Alnus serrulata)
A spreading shrub or tree up to 15 ft. tall with smooth, thin, gray bark. Flowers small, imperfect, the male catkin 2-4 in. long, the female "cone" ¾ in. long. Fruit a woody cone ¾ in. long. It grows in bogs, swamps and along streams in eastern Oklahoma and Texas and Arkansas and Louisiana. It belongs to the birch family *(Betulaceae)*.

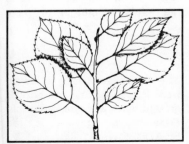

Fig. 335

SHADBLOW
(Amelanchier arborea)
A shrub with erect, parallel branches or a tree up to 60 ft. tall. Flowers white, fragrant, 2 in. across. Fruit a reddish, dry apple (pome) about ½ in. in diameter. It grows on rich wooded slopes or along streams in Louisiana, Arkansas, eastern Oklahoma and northeastern Texas. It belongs to the rose family *(Rosaceae)*.

334

1. If the leaves range in length from 4 to 7 in., it is the SOUR-WOOD.

KI 168

2. If the leaves are 4 in. or less long, as symbolized in (KI 169), to the right, find this picture in the left-hand margin of page 337.

335

Fig. 336

SOURWOOD

(Oxydendrum arboreum)

A pendulous-branched tree up to 70 ft. tall. Flowers white, fragrant, cylindric, ⅓ in. long, borne in terminal clusters 6-12 in. long. The fruit is a dry capsule ½ in. long. It grows on wooded slopes in Arkansas and eastern Louisiana. It belongs to the heath family *(Ericaceae)*.

> 4"

KI 169

1. If the young twigs are yellow-green, it is the MOUNTAIN WILLOW.

2. If the young twigs are reddish-brown and the bark is divided into irregularly shaped scales, it is the CHOKE-CHERRY.

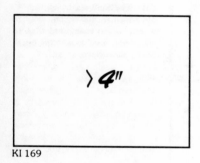

KI 169

3. If the young twigs are reddish-brown and the bark is divided into long plates, it is the AMERICAN PLUM.

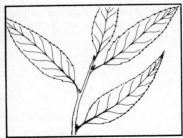

Fig. 337

MOUNTAIN WILLOW
(Salix monticola)
Shrub up to 18 ft. tall. Flowers imperfect, the catkins 2 in. long. The fruit is a tiny capsule, bearing cottony seeds. It grows in wet meadows or along streams in the mountains of New Mexico. It is in the willow family *(Salicaceae)*.

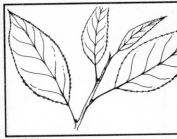

Fig. 338

CHOKECHERRY
(Prunus virginiana)
A large shrub to small tree up to 30 ft. tall with horizontal branches. The flowers are ⅓ in. across, white and borne in elongate clusters 3-6 in. long. The fruit is a cherry, red or dark red, and ⅓ in. in diameter. It grows on open slopes, around rimrock, in swamps or breaks throughout most of our 5-state area. It belongs to the rose family *(Rosaceae)*.

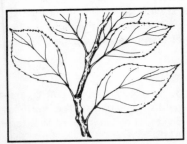

Fig. 339

AMERICAN PLUM
(Prunus americana)
A thicket-forming shrub or small tree up to 35 ft. tall. The flower is white, 1 in. across and many-stamened. The fruit is a red to yellow plum, 1 in. long. It grows in rich soil in Oklahoma, Arkansas and Louisiana. It belongs to the rose family *(Rosaceae)*.

1. If the outline of some or all of the leaves of your specimen, is reverse tear-, egg-, or triangle-shape, as illustrated in (KI 171), to the right, go to this picture in the left-hand margin of page 341.

KI 170

2. If the outline of the leaves is tear-, egg-, diamond-, triangle-shape or the margins are parallel, go to the diagram (KI 176), shown to the right in the left-hand margin of page 351.

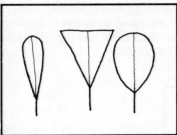

KI 171

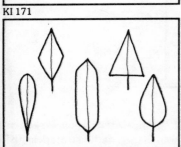

KI 176

1. If the leaf is reverse triangular in outline, it is the BLACK JACK OAK.

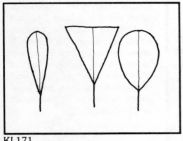

KI 171

2. If the leaf is reverse egg-or tear-shape, as illustrated in (KI 172), to the right, find this drawing in the left-hand margin of page 343.

341

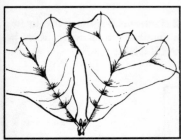

Fig. 340

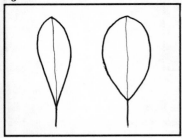

KI 172

BLACKJACK OAK

(Quercus marilandica)

A small tree up to 30 ft. tall with very rough, black, furrowed bark. The tiny, imperfect flowers are borne in separate clusters. The male lax spike (catkin) is conspicuous, hairy, yellowish-green and 2-4 in. long. The fruit is an acorn (nut with a basal, woody cup) about 2/5 in. long. It grows on sandy soil in upland forests in the eastern half of Oklahoma and Texas and throughout Arkansas and Louisiana. It belongs to the beech family *(Fagaceae)*.

342

1. If the leaf is 5-9 lobed, go to (KI 174), shown to the right, in the left-hand margin of page 347.

2. If the leaf is unlobed but has blunted marginal teeth, it is the CHESTNUT OAK.

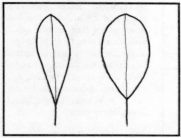

KI 172

3. If the leaf is unlobed but has sharp marginal teeth, look for (KI 173), at the right, in the left-hand margin of page 345.

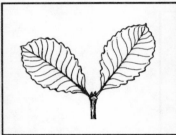

KI 174

5-9 lbd

Fig. 341

CHESTNUT OAK
(Quercus prinus)
A large tree up to 100 ft. tall. The tiny, imperfect flowers are borne in separate clusters. The male lax spike (catkin) is conspicuous, slender, green, hairy and 3-4 in. long. The fruit is an acorn (nut with a basal cup) about 1-1-½ in. long. It grows in moist forests in eastern Oklahoma and Texas and Louisiana and Arkansas. It belongs to the beech family *(Fagaceae)*.

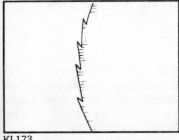

KI 173

1. If the leaf is few-toothed in the manner shown in the diagram to the right, it is NET-LEAF OAK.

2. If the length of the leaf is 4-7 in., it is SWAMP OAK.

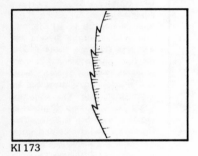

KI 173

3. If the length of the leaf is 1-4 in., and your specimen is common west of the Pecos River (Texas) into New Mexico, it is the ARIZONA OAK.

4. If the length of the leaf is 1-4 in. and your specimen occurs on sandy acid soils in east Texas and Arkansas, it is RHODODENDRON.

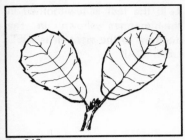

Fig. 342

Fig. 343

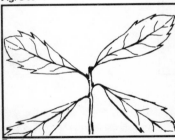

Fig. 344

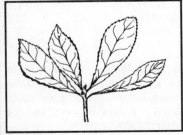

Fig. 345

NET-LEAF OAK
(Quercus rugosa)
A shrub but usually a tree up to 40 ft. tall. Flowers tiny, imperfect, the male catkin greenish-yellow 1-2 in. long. Fruit an acorn 2/5-4/5 in. long. A rare tree in New Mexico and far western Texas. It is in the beech family *(Fagaceae)*.

SWAMP OAK
(Quercus bicolor)
Pendulous-branched tree up to 100 ft. tall. Flowers tiny, imperfect, the male catkin yellowish-green, 2-4 in. long. Fruit an acorn about ¾-1-¼ in. long. It grows in swampy habitats in Arkansas and eastern Oklahoma. It belongs to the beech family *(Fagaceae)*.

ARIZONA OAK
(Quercus arizonica)
A montane shrub at high elevation or a tree up to 40 ft. tall on the lower slopes. The male lax spike (catkin) is pale yellow, conspicuous and 1-3 in. long. The fruit is an acorn ⅓-⅔ in. long. It occurs in the mountains of New Mexico and far western Texas. It belongs to the beech family *(Fagaceae)*.

RHODODENDRON
(Rhododendron spp.)
A group of vegetatively similar shrubs 1-20 ft. tall. Flower showy, pink to white, trumpet-shaped with protruding, hairy stamens (if included, then called AZALEAS), ranging from 1¼ to 1¾ in. long. Fruit a dry capsule ⅓-¾ in. long. They grow in acid, wet soils in Arkansas, Louisiana and eastern Texas. They are in the heath family *(Ericaceae)*.

1. If the blade of a leaf is lobed in the form of a cross as depicted to the right, it is POST OAK.

2. If the leaves are not cross-shaped and range in length from 2 to 6 in., it is GAMBEL OAK.

5-9 lbd

KI 174

3. If the longest leaves are 10-12 in. long, and not cross-shaped, go to (KI 175), shown to the right, in the left-hand margin of page 349.

Fig. 346

POST OAK
(Quercus stellata)
Stout-limbed tree up to 70 ft. tall. Flowers tiny, imperfect, the male catkin, yellow, hairy, 2-4 in. long. Fruit an acorn, ½-¾ in. long. It is the predominant tree of the Texas Cross Timbers and Oak Forest of central Oklahoma, ranging east and north to Arkansas and Louisiana and south in Texas. It belongs to the beech family *(Fagaceae)*

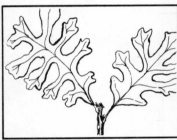

Fig. 347

GAMBEL OAK
(Quercus gambelii)
Mostly a thicket-forming shrub or tree up to 45 ft. tall. Flowers tiny, imperfect, the male catkin reddish-brown, hairy, 1-1½ in. long. Fruit an acorn ½-¾ in. long. It grows at mid to high elevations in the mountains of far western Texas and New Mexico. It belongs to the beech family *(Fagaceae)*.

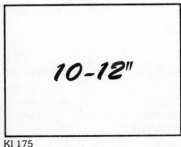

10-12"

KI 175

348

1. If the lobes are shallow in the manner shown to the right, it is OVERCUP OAK.

10-12"

KI 175

2. If the lobes are deeply cut, as shown to the right, it is BUR OAK.

Fig. 348

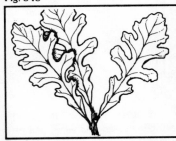
Fig. 349

OVERCUP OAK
(Quercus lyrata)
An open, irregularly crowned tree of moderate height. The tiny, imperfect flowers are borne in separate clusters, the male lax spike (catkin) is hairy, slender, yellow and 3-6 in. long. The fruit is an acorn (nut with a basal, woody cup) in which the cup often encloses the nut and is 1/20-1 in. long and 1 in. in diameter. It grows in streamside forests in Arkansas, eastern Oklahoma and Texas and Louisiana. It is in the beech family (Fagaceae).

BUR OAK
(Quercus macrocarpa)
A massive tree up to 150 ft. tall. The tiny, imperfect flowers are borne in separate clusters. The male lax spike (catkin) is conspicuous, yellowish-green and 4-6 in. long. The fruit is an acorn (nut with a basal, woody cup) that is quite large, up to 2 in. long. The cup is fringed on its rim and encloses about ¾ of the nut. It grows along streams in the northern half of central and eastern Texas, the northern half of Louisiana, the eastern half of Oklahoma and through Arkansas. It belongs to the beech family (Fagaceae).

350

1. If most or all of the leaves are tear-shaped, or are narrow with parallel margins, proceed to (KI 177), as shown to the right, in the left-hand margin of page 353.

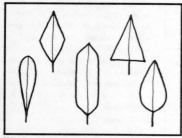

KI 176

2. If the leaves are egg-, triangle-, wedge-shape, elliptic in outline or the blade is broad with parallel sides, go to the diagram (KI 181), as shown to the right, in the left-hand margin of page 361.

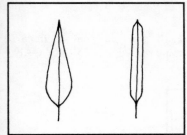

KI 177

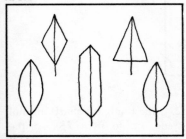

KI 181

1. If the marginal teeth are spiny-tipped, it is EMORY OAK.

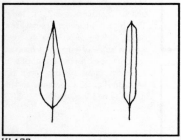

KI 177

2. If the teeth are sharp but not spiny, go to (KI 178), at the right, in the left-hand margin of page 355.

Fig. 350

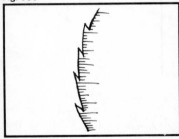

KI 178

EMORY OAK
(Quercus emoryi)
A small or large tree up to 60 ft. tall with deeply furrowed, black bark. The tiny, imperfect flowers are borne in separate clusters. The male cluster is conspicuous, being a lax spike (catkin) that is yellow, hairy and 1-2 in. long. The fruit is an acorn (nut with a basal, woody cup) ½-¾ in. long with a shallow cup. It grows on rocky, upper mountain slopes in southern New Mexico and far western Texas. It belongs to the beech family *(Fagaceae)*.

1. If the older leaves are hairy on the under-surface veins only, it is BLACK WILLOW.

2. If the upper and lower blade surfaces are densely hairy, it is HEART-LEAF WILLOW.

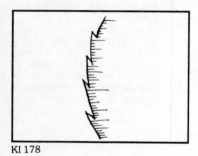

KI 178

3. If only the undersurface of the blade, the veins as well, is hairy, go (KI 179), at the right, in the left-hand margin of page 357.

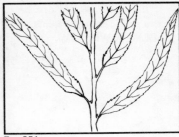

Fig. 351

BLACK WILLOW
(Salix nigra)
A tree up to 80 ft. tall or more. The tiny, imperfect flowers are borne in lax spikes (catkins) that are conspicuous, slender and 1-2 in. long. The fruit is a tiny capsule bearing cottony seeds. It grows in wet habitats in Arkansas, Louisiana, eastern Oklahoma and the eastern ⅔ of Texas. It belongs to the willow family *(Salicaceae).*

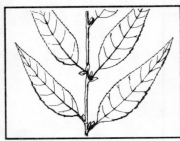

Fig. 352

HEART-LEAF WILLOW
(Salix cordata)
A shrub 6-12 ft. tall. Flowers, tiny, imperfect, the catkins 2/5-1 in. long. The fruit is a tiny capsule bearing cottony seeds. It grows in wet sands of dunes or lake beaches in Arkansas. It belongs to the willow family *(Salicaceae).*

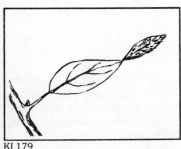

KI 179

356

1. If the blade is remotely toothed, as shown to the right, your specimen is SWEETLEAF.

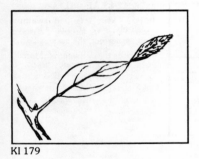

KI 179

2. If the blade is regularly toothed, as drawn in (KI 180), to the right, go to this drawing in the left-hand margin of page 359.

357

Fig. 353

KI 180

SWEETLEAF

(Symplocos tinctoria)

A tree or shrub up to 20 ft. tall. Flowers yellowish-white, ⅓ in. long with many protruding stamens. The fruit is dry, olive-like, orangish-brown and about ½ in. long. It grows in streamside forests, bay flats and swampy ground in Arkansas, eastern Oklahoma, Texas and Louisiana. It belongs to the sweet-leaf family *(Symplocaceae)*.

1. If the margin is coarse-toothed, as shown to the right, it is CHINQUAPIN OAK.

2. If the marginal teeth are fine and the blade base is narrow, it is SANDBAR WILLOW.

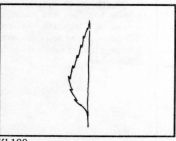

KI 180

3. If the marginal teeth are fine and the blade is broad at the base, it is WILD PLUM.

Fig. 354

CHINQUAPIN OAK
(Quercus muhlenbergii and/or *prinoides)*
Both species have similar foliage and the same common name. The latter is a shrub or small tree up to 15 ft. tall. The former is a tree up to 60 ft. tall. Their tiny imperfect flowers are borne in separate clusters of which the male lax spike (catkin) is conspicuous, yellow and 1-4 in. long. The fruit is an acorn (nut with a woody, basal cup) ½-1 in. long. The shrub species grows in sunny, rocky sites in Arkansas and eastern Oklahoma and the tree species grows in uplands of Arkansas, Oklahoma and Louisiana and in the northern half of Texas. They belong to the beech family *(Fagaceae)*.

Fig. 355

SANDBAR WILLOW
(Salix interior)
A thicket-forming shrub rarely more than 12 ft. tall. Flowers inconspicuous, in catkins ¾-3 in. long. The fruit is a small capsule. It is wide-ranging in our 5-state area along streams and around lakes. It belongs to the willow family *(Salicaceae)*.

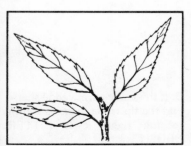

Fig. 356

WILD PLUM
(Prunus spp.)
Several low, thicket-forming, shrubby species. The flowers are white, about ½ in. across, bearing many stamens. The fruit is a small, yellow to red plum from ½ to 1 in. long. They grow mostly in moist sites throughout our 5-state area. They belong to the rose family *(Rosaceae)*.

1. If the leaves are broadly triangular in outline, as depicted in (KI 182), to the right, find this illustration in the left-hand margin of page 363.

2. If the leaves are egg-shape to elliptic, or broad with parallel sides, find the diagram to the right (KI 183), in the left-hand margin of page 365.

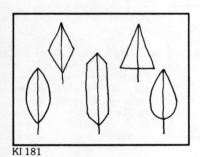

KI 181

3. If the leaves are wedge-shape and the tip tri-toothed, as shown to the right, it is BITTER BRUSH.

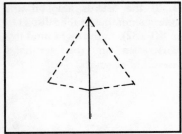

KI 182

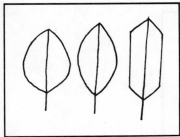

KI 183

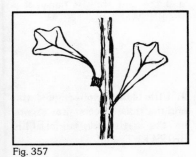

Fig. 357

BITTER BRUSH

(Purshia tridentata)

A prostrate to erect shrub up to 10 ft. tall. The yellow flowers are many-stamened and ⅓ in. across. The fruit is slender, seed-like and about ½ in. long. It grows in the Tunitcha Mountains of New Mexico on sunny slopes. It belongs to the rose family *(Rosaceae)*.

362

1. If the leaf is unlobed and hairy beneath, it is the BIRCH.

2. If the leaf is lobed, the lobes are entire and the blade hairy on both surfaces, it is the ARIZONA PLANE-TREE.

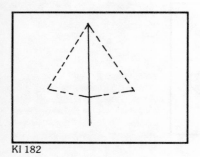

KI 182

3. If the leaf is lobed, only the undersurface veins are hairy and the leaf stalk is hairy, it is the SYCAMORE.

4. If the leaf is lobed, only the undersurface veins are hairy and the leaf stalk is hairless, it is the SWEETGUM.

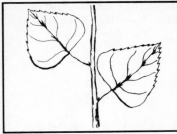
Fig. 358

BIRCH
(Betula nigra)
A tree up to 90 ft. tall with peeling, pinkish bark. The flowers are borne in separate lax spikes (catkins). The male catkin is brownish and 1-3 in. long. The female is green and ½ in. long. It grows in Louisiana, Arkansas, eastern Oklahoma and eastern Texas. It belongs to the birch family *(Betulaceae)*.

Fig. 359

ARIZONA PLANE-TREE
(Platanus wrightii)
A tree up to 80 ft. tall with bark that peels off in patches. The flowers are borne in spherical heads that enlarge in fruit to ¾ in. diameter. It grows along streams in southwestern New Mexico. It belongs to the sycamore family *(Platanaceae)*.

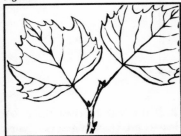
Fig. 360

SYCAMORE
(Platanus occidentalis)
A tree up to 150 ft. tall with mottled bark as a patterned patchwork of brown, green, white and buff. Flowers borne in spheres 1 to 1½ in. in diameter in fruit. It grows in all of our area except far western Texas and New Mexico. It belongs to the sycamore family *(Platanaceae)*.

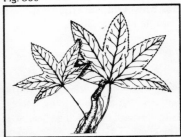
Fig. 361

SWEETGUM
(Liquidambar styraciflua)
A tree up to 120 ft. tall with corky-winged twigs. Flowers borne in globose clusters that enlarge in fruit to 1-1½ in. in diameter. It grows in Arkansas, Louisiana, eastern Oklahoma and Texas. It is in the witch hazel family *(Hamamelidaceae)*.

364

1. Check the leaf-stalks of several *older* leaves, if they are somewhat to very hairy, go to (KI 184), shown to the right, in the left-hand margin of page 367.

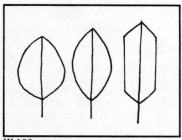

KI 183

2. If the leaf-stalks of the *older* leaves are smooth and non-hairy (glabrous), find (KI 193), shown to the right, in the left-hand margin of page 385.

stalks hairy

KI 184

stalks non-hairy

KI 193

1. If *any* of the leaves are lobed, as shown to the right, it is SHIN OAK.

2. If the marginal teeth are stiff-spiny to the touch, it is WESTERN SCRUB OAK.

stalks hairy

KI 184

3. If the marginal teeth are sharp or blunted but not spiny, find (KI 185), to the right, in the left-hand margin of page 369.

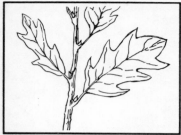

Fig. 362

SHIN OAK
(Quercus havardii and *undulata)*
Low shrubs, often in thickets. Flowers tiny, imperfect, the male catkin ½-1½ in. long. The fruit is an acorn ½-1 in. long. Of these similar oaks, *Q. havardii* grows in sand in the lower panhandle of Texas into eastern New Mexico and *Q. undulata* on rocky slopes at upper altitudes in the mountains of the Texas spur north into New Mexico. They belong to the beech family *(Fagaceae).*

Fig. 363

WESTERN SCRUB OAK
(Quercus turbinella)
An evergreen shrub up to 12 ft. tall. Flowers tiny, imperfect, the male catkin yellowish-green, ¼-¾ in. long. The fruit is an acorn 3/5 in. long. It grows on dry lower slopes in the mountains of New Mexico and far western Texas. It belongs to the beech family *(Fagaceae).*

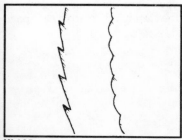

KI 185

1. If the marginal teeth are double saw-toothed, as shown in the diagram (KI 186), to the right, go to this diagram in the left-hand margin of page 371.

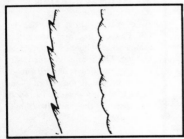

KI 185

2. If the marginal teeth are a single series of blunted or sharp teeth like the diagram (KI 89), to your right, find this diagram, in the left-hand margin of page 377.

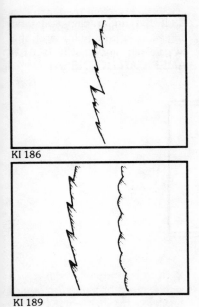

KI 186

KI 189

1. If the underleaf surface hairiness is found only on the veins or vein angles and the twigs are corky-winged, it is WINGED ELM.

2. If the underveins and vein angles only are hairy and the twigs lack corky wings, it is AMERICAN HORNBEAM.

KI 186

3. If any of the underblade surface is hairy, as depicted in (KI 187), to the right, go to this drawing in the left-hand margin of page 373.

Fig. 364

WINGED ELM
(Ulmus alata)
A tree up to 45 ft. tall with usually prominent, corky-winged twigs. Flowers inconspicuous, the fruit seed-like, winged, green and ¼-½ in. long. It grows in upland forests in the eastern half of Oklahoma, Texas, Louisiana and Arkansas. It is in the elm family *(Ulmaceae)*.

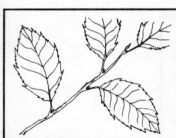
Fig. 365

AMERICAN HORNBEAM
(Carpinus caroliniana)
A small tree up to 30 ft. tall with a fluted, flattened trunk and smooth, gray bark. The small flowers are imperfect and borne in separate lax clusters (catkins). The male catkin is green, slender, elongate from 1 to 1½ in. long. The female catkin is 2 in. long with prominent, arrowhead-shaped bracts. The fruit is a nutlet about ⅓ in. long. It grows in rich woods and on bottom lands in Arkansas, Louisiana, eastern Oklahoma and Texas. It belongs to the birch family *(Betulaceae)*.

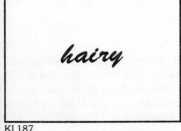

hairy

KI 187

372

1. If both blade surfaces are hairy and the leaves are 2-3 in. long, it is MEXICAN ALDER.

2. If only the underblade surface is hairy and the longest leaves are 2 in. or less, it is CEDAR ELM.

hairy

KI 187

3. If only the underblade surface is hairy, and the longest leaves are between 2 and 6 in. long, go to (KI 188), at the right, in the left-hand margin of page 375.

Fig. 366

MEXICAN ALDER
(Alnus oblongifolia)
A shrub or tree up to 30 ft. tall. Flowers tiny, imperfect, the male catkin yellowish-orange, 2 in. long. Female cluster a "cone" and the fruit a nutlet within. It grows along streams in the lower reaches of the mountains in western New Mexico. It is in the birch family *(Betulaceae)*.

Fig. 367

CEDAR ELM
(Ulmus crassifolia)
A tree up to 75 ft. tall with gray, flat-ridged bark. The small flowers lack petals and are inconspicuous. The seed-like fruit is oblong, green winged and ¼-½ in. long. It grows on floodplains or in upland forests in Arkansas, Louisiana, and the eastern half of Oklahoma and Texas. It belongs to the elm family *(Ulmaceae)*.

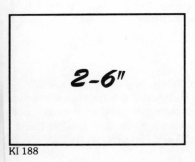

2-6"

KI 188

374

1. If the leaf-tip angle is narrow, as shown to the right, it is the HAZELNUT.

2. If the leaf-tip is broad and some of the twigs corky-winged, it is ROCK ELM.

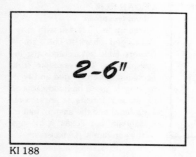

2-6"

KI 188

3. If the leaf-tip is broad, none of the twigs are winged, and the twigs are hairy, it is WOOLY HORNBEAM.

4. If the leaf-tip is broad and the twigs are neither winged nor hairy, it is the BIRCH-LEAF BUCKTHORN.

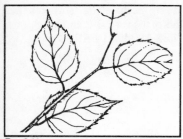

Fig. 368

Fig. 369

Fig. 370

Fig. 371

HAZELNUT
(Corylus americana)
A thicket-forming shrub 3-10 ft. tall. The imperfect flowers are borne in separate clusters. The male pendulous spike (catkin) is slender and 3-4 in. long. The female flowers are enclosed by two reddish-brown bracts which are leaf-like and also sheath the fruit, a globose, brown nut about ½ in. in diameter. It grows in upland forests in Arkansas and eastern Oklahoma. It belongs to the birch family *(Betulaceae)*.

ROCK ELM
(Ulmus thomasi)
A clean-trunked tree up to 100 ft. tall. Flower inconspicuous, the seed-like fruit winged, densely hairy and ½ in. long. It grows on rocky-gravelly riverbanks and slopes in northwestern Arkansas. It is in the elm family *(Ulmaceae)*.

WOOLY HORNBEAM
(Ostrya virginiana)
A tree up to 60 ft. tall. Flowers imperfect, the male catkin green to red 1½-3 in. long, the female catkin enlarging into a "cone" in fruit, 1½-2 in. long. It grows in rich, upland forests in Arkansas, Louisiana, Oklahoma and eastern Texas. It is in the birch family *(Betulaceae)*.

BIRCH-LEAF BUCKTHORN
(Rhamnus betulaefolia)
A shrub or small tree up to 20 ft. tall. The flower is minute and greenish. The fruit is olive-like (drupe), blackish, globose and ⅓ in. in diameter. It grows in moist canyons in New Mexico and far western Texas. It belongs to the buckthorn family *(Rhamnaceae)*.

1. If both blade surfaces are hairy (leaves 1-2 in. long), it is TEXAS MULBERRY.

2. If only the undersurface veins are hairy, it is OPPOSUM WOOD.

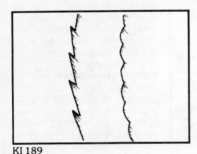

KI 189

3. If the blade undersurface is hairy, go to (KI 190), shown to the right, in the left-hand margin of page 379.

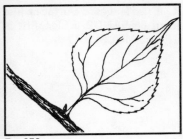

Fig. 372

TEXAS MULBERRY
(Morus microphylla)
A shrub or shrubby tree up to 20 ft. tall. Flowers imperfect, inconspicuous, the female catkin ripening into a red to black "berry," the male catkin green to reddish, ½-¾ in. long. It grows on dry, rocky, limestone hills in New Mexico and the western ⅔ of Texas. It belongs to the Mulberry family *(Moraceae)*.

Fig. 373

OPPOSUM WOOD
(Halesia carolina)
A large shrub or tree up to 35 ft. tall. The flower is an open, white, 4-lobed bell about ¾ in. across. The fruit is dry-fleshy with four large wings and ranges 1-2 in. in length. It grows in protected woods or along streams in Arkansas, Oklahoma, northern Louisiana and northeastern Texas. It belongs to the storax family *(Styraceae)*.

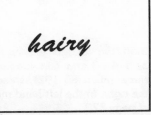

hairy

KI 190

378

1. If the leaves are 1 in. or less long, it is SANDPAPER OAK.

2. Measure the longest leaves on your specimen, if they range from 1 to 3 in. long, go to (KI 191), shown to the right, in the left-hand margin of page 381.

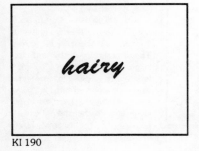

KI 190

3. If the longest leaves range between 3 and 7 in. long, go to the symbol (KI 192), shown at the right, in the left-hand margin of page 383.

Fig. 374

SANDPAPER OAK

(Quercus pungens)

An evergreen shrub or small tree. Flowers tiny, imperfect, the male catkin 1½ in. long. Fruit an acorn about 2/5 in. long. It grows on dry, lower mountain slopes in western Texas and New Mexico. It belongs to the beech family *(Fagaceae).*

1-3"

KI 191

3-7"

KI 192

1. If the marginal teeth are barely evident, it is LEUCOTHAE.

2. If the marginal teeth are evident and glandular, it is DOGBERRY.

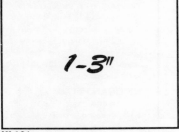

KI 191

3. If the marginal teeth are evident, non-glandular and the blade is 1-2 in. wide, it is MEXICAN PLUM.

4. If the marginal teeth are evident, non-glandular and the blade is ½-1 in. wide, it is SWEETSPIRE.

Fig. 375

Fig. 376

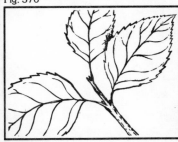

Fig. 377

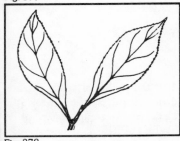

Fig. 378

LEUCOTHAE
(Leucothoe elongata and *racemosa)*
Shrubs 3-12 ft. tall, differing mostly by a single flower characteristic, sepal length. Flower cylindric, white to pink 2/5 in. long. Fruit a rounded capsule 1/5 in. in diameter. They grow in wet sand or acid swamps in Louisiana. They belong to the heath family *(Ericaceae)*.

DOGBERRY
(Pyrus arbutifolia)
A shrub up to 12 ft. tall, forming thickets or a colony of small trees. The pinkish-white flowers have many stamens and are about ½ in. across, occurring in showy clusters about 2 in. across. The fruit is a small, red apple (pome) about ¼ in. in diameter. It grows in swamps, bogs and wet ground in eastern Texas and Oklahoma, Arkansas and Louisiana. It belongs to the rose family *(Rosaceae)*.

MEXICAN PLUM
(Prunus mexicana)
A shrub or small tree up to 25 ft. tall. Flower, white, showy, many-stamened, 1 in. across. The fruit is a purple plum 1 in. in diameter. It grows in upland and streamside forests in Louisiana, Arkansas, eastern Oklahoma, central and northeastern Texas. It is in the rose family *(Rosaceae)*.

SWEET SPIRE
(Itea virginica)
See page 304.

382

1. If the bark is smooth and gray, it is the RED BEECH.

2. If the bark is fissured, brownish and the leaves are 1-5 in. wide, it is the SWAMP COTTONWOOD.

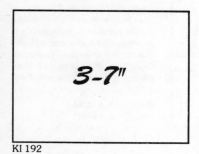

KI 192

3. If the bark is fissured, brownish and the leaves are 2-3 in. wide, it is the SNOW-DROP TREE.

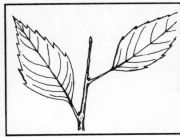
Fig. 379

RED BEECH
(Fagus grandifolia)
Large tree up to more than 100 ft. tall with smooth, gray bark. Flowers tiny, imperfect, the male catkin globose, green, 1 in. in diameter. The fruit is bur-like, ¾ in. long. It grows in moist forests in eastern Oklahoma, Texas, Arkansas and Louisiana. It belongs to the beech family *(Fagaceae)*.

Fig. 380

SWAMP COTTONWOOD
(Populus heterophylla)
Tree up to 90 ft. tall. Flowers tiny, imperfect, the catkins 2-6 in. long. The fruit is a small capsule, bearing cottony seeds. It grows around swamps and near streams in Louisiana and Arkansas. It is in the willow family *(Salicaceae)*.

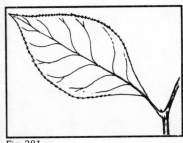
Fig. 381

SNOWDROP TREE
(Halesia diptera)
A shrub or tree up to 25 ft. tall. Flowers drooping, white, bell-shaped and about 1 in. long. The dry fruit is oblong, conspicuously 2-winged and 1-2 in. long. It grows near water in eastern Oklahoma, Texas, Arkansas and Louisiana. It belongs to the storax family *(Styracaceae)*.

1. If the longest leaves are 1 in. or less long and your specimen occurs in the Big Bend region of Texas, it is BIG BEND SERVICE-BERRY.

2. If the longest leaves are 1 in. or less long and your specimen occurs in the western half of New Mexico, it is MOUNTAIN MAHOGANY.

stalks

non-hairy

KI 193

3. If the longest leaves exceed 1 in., go to (KI 194), at the right, in the left-hand margin of page 387.

385

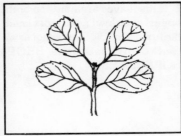

Fig. 382

BIG BEND SERVICE-BERRY
(Amelanchier denticulata)
A shrub 3-9 ft. tall with dense foliage. The white flower is ½ in. across with many stamens. The fruit is a small, dark apple (pome) ½ in. in diameter. It grows around rimrock or on rocky slopes in the Big Bend mountains of Texas. It belongs to the rose family *(Rosaceae)*.

Fig. 383

MOUNTAIN MAHOGANY
(Cercocarpus montanus)
Shrub or small tree 3-18 ft. tall. The silky flower is tubular ⅓ in. long with no petals but many stamens. The fruit is a silky plume up to 2½ in. long. It grows in arid habitats from central Texas north into New Mexico. It belongs to the rose family *(Rosaceae)*.

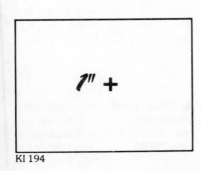

1″ +

KI 194

1. If the leaves are hairy on both upper and lower surfaces and the margin is double saw-toothed, it is the HOP HORNBEAM.

2. If the leaves are hairy on both surfaces, the margin has one series of saw-teeth and the leaves are 3 in. or less long, it is the OKLAHOMA ALDER.

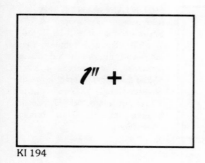

KI 194

3. If the leaves are hairy on both surfaces, the margin has one series of saw-teeth and the leaves are 6 in. or less long, it is the COASTAL LEUCOTHOE.

4. If the leaves are hairy only beneath, go to (KI 195), at the right, in the left-hand margin of page 389.

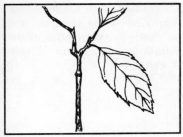

Fig. 384

HOP HORNBEAM
(Ostrya knowltonii)
Small tree up to 35 ft. tall.
Flowers tiny, imperfect, the male
catkin yellowish-green to brown,
about 1 in. long, the female
catkin cone-like, ripening into a
fruit ¾ in. broad and 1½ in. long.
It grows in mountain canyons of
New Mexico and western Texas.
It belongs to the birch family
(Betulaceae).

OKLAHOMA ALDER
See page 296.

Fig. 385

COASTAL LEUCOTHOE
See page 304.

Fig. 386

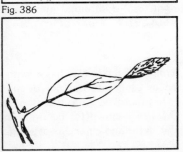

KI 195

388

1. If some of the leaves are deeply lobed and some are not, as shown to the right, and the margin has two series of saw-teeth, it is the RED MULBERRY.

2. If some of the leaves are lobed and some are not, as shown to the right, and the margin has one series of saw-teeth, it is PAPER MULBERRY.

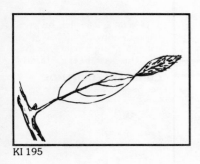

KI 195

3. If all of the leaves are deeply lobed, go to (KI 196), at the right, in the left-hand margin of page 391.

4. If all of the leaves have toothed or scalloped but not deeply lobed margins, go to this symbol (KI 199), shown to the right, in the left-hand margin of page 297.

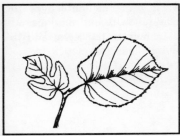
Fig. 387

RED MULBERRY
(Morus rubra)
Tree up to 70 ft. tall. Flowers tiny, imperfect, the sexes on separate trees. Male catkins green 2-3 in. long. Female spike ripening into a multiple fruit ("berry") that is black and about 1 in. long. It grows in rich, moist soil in Texas, Oklahoma, Louisiana and Arkansas. It belongs to the mulberry family *(Moraceae)*.

Fig. 388

PAPER MULBERRY
(Broussonetia papyrifera)
A small tree up to 50 ft. tall. Flowers inconspicuous, imperfect, the sexes on separate trees. Male catkins 2½-3½ in. long. Female spikes globose, ripening into a reddish multiple fruit ¾ in. in diameter. An occasional escape from cultivation throughout our 5-state area. It is in the mulberry family *(Moraceae)*.

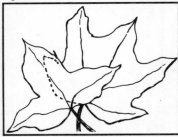
KI 196

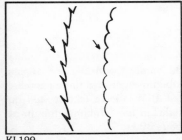
KI 199

1. If the undersurface of the blade is hairy and the lobes are 7 in number, it is the BLACK OAK.

2. If the undersurface of the blade is hairy and the lobes are 3 to 11 in number, it is RED OAK.

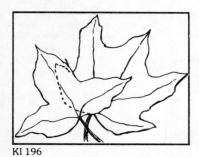

KI 196

3. If only the undervein angles are hairy with the blade hairless, go to (KI 197), shown to the right, in the left-hand margin of page 393.

BLACK OAK
See page 312.

Fig. 389

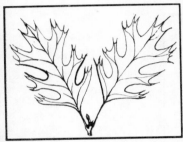

RED OAK
See page 312.

Fig. 390

hairy

KI 197

1. If the branchlets are red, it is the TEXAS RED OAK.

2. If the branchlets are gray to brown and the lobes are shallow cut toward the mid-rib, it is SHUMARD'S OAK.

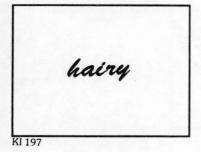

KI 197

3. If the branchlets are green to brown and the lobes are deep-cut toward the midrib, go to (KI 198), shown at the right, in the left-hand margin of page 395.

Fig. 391

TEXAS RED OAK
(Quercus texana)
A small tree up to 30 ft. tall with furrowed black or smooth, gray bark. Flowers tiny, imperfect, the male catkin 1½-3½ in. long. The fruit is an acorn up to ⅓ in. long. It grows on limestone slopes in central Texas and southern Oklahoma in the Arbuckle Mountains. It is in the beech family *(Fagaceae)*.

Fig. 392

SHUMARD'S OAK
See page 312.

green-brown

KI 198

394

1. If the lobe cuts are angled, as shown to the right, it is the PINOAK.

2. If most of the lobe cuts are rounded, as shown to the right and your specimen is found in the sterile upland sands of Eastern Oklahoma, it is the SCARLET OAK.

green-brown

KI 198

3. If the lobe cuts are rounded and your specimen is found in streamside forests, it is the NUT-TALL OAK.

Fig. 393

PINOAK
(Quercus palustris)
A large tree up to 120 ft. tall with many short branches. Flowers tiny, imperfect, the male catkin 2-3 in. long. The fruit is an acorn ½ in. in diameter. It grows in bottomland forests in Arkansas and eastern Oklahoma. It is in the beech family *(Fagaceae)*.

Fig. 394

SCARLET OAK
(Quercus coccinea)
Tree up to 100 ft. tall. Flowers tiny, imperfect, the male catkin reddish, 3-4 in. long. The fruit is an acorn up to 1 in. long. It grows on sterile sands in eastern Oklahoma. It is in the beech family *(Fagaceae)*.

Fig. 395

NUTTALL OAK
(Quercus nuttallii)
A tree up to 120 ft. tall, the trunk becoming buttressed with age. Flowers tiny, imperfect, the male catkins and acorns like those of RED OAK. It grows in bottomland, wet forests in Louisiana, Arkansas and eastern Oklahoma. It belongs to the beech family *(Fagaceae)*.

396

1. If the margin is scalloped, it is the OZARK WITCH-HAZEL.

2. If the margin is bristle-toothed, it is the CHINGUAPIN.

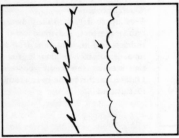

KI 199

3. If the margin is once- or twice-toothed, go to (KI 200), shown to the right, in the left-hand margin of page 399.

Fig. 396

OZARK WITCH-HAZEL

See WITCH-HAZEL on page 290.

Fig. 397

CHINQUAPIN

(Castanea spp.)

A group of shrubs to small trees imperfectly separated into species and or varieties at this time. In addition to their distinctive leaves, they have tiny, imperfect flowers borne in separate clusters, of which the male lax spike (catkin) is hairy, slender and 2-8 in. long. The female cluster ripens into bur fruits with long prickles. Each is ¾ in. in diameter enclosing brown, angled nuts. They grow in open forests and thickets in Louisiana, Arkansas, in eastern Oklahoma and Texas. They belong to the beech family *(Fagaceae)*.

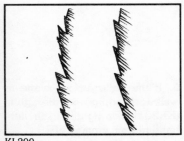

KI 200

1. If the undersurface midvein or veins only are hairy and the leaf is toothed at the base, it is BLACK CHERRY.

2. If the undersurface veins of the older leaves are hairy and the leaf is not toothed at the base, it is MONTANE SILVERBELL.

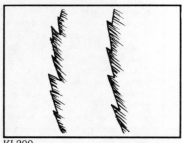

KI 200

3. If the undersurface blade as well as the veins are hairy, go to (KI 201), to the right, in the left-hand margin of page 401.

Fig. 398

BLACK CHERRY
(Prunus serotina)
A tree up to 90 ft. tall with reddish-brown, aromatic bark. The white flowers have numerous stamens and are about ¼ in. across. The fruit is a black cherry ⅓-½ in. in diameter. It grows throughout Texas, Oklahoma, Arkansas and Louisiana. It is in the rose family *(Rosaceae)*.

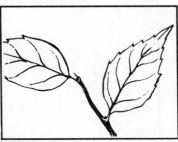

Fig. 399

MONTANE SILVERBELL
(Halesia monticola)
A tree up to 90 ft. tall. Flower white, bell-shaped and about 1 in. across. The fruit is a dry capsule with four wings and is 2 in. long. It grows in the mountainous regions of Oklahoma and Arkansas. It belongs to the storax family *(Styracaceae)*.

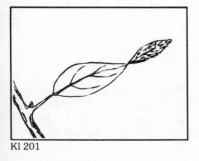

KI 201

400

1. If any of the leaves are curved, as shown to the right, it is HACKBERRY.

2. If the leaves are straight and double-toothed, it is MOUN-TAIN-SPRAY.

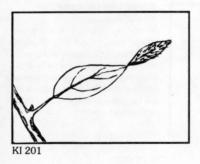

KI 201

3. If the leaves are straight and single-toothed, go to (KI 202), shown to the right, in the left-hand margin of page 403.

HACKBERRY
See page 318.

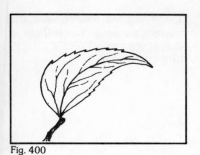

Fig. 400

MOUNTAIN-SPRAY
(Holodiscus discolor)
Mostly a 3 ft. tall shrub occasionally up to 12 ft. tall with branches to the base and shreddy bark. The small, white to pink flowers are borne in showy, terminal clusters 2-8 in. long. The fruit is very small, dry, splitting along one suture (follicle). It grows in the mountains of New Mexico and far western Texas. It belongs to the rose family *(Rosaceae)*.

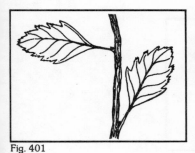

Fig. 401

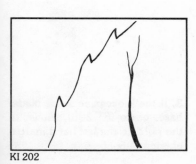

KI 202

1. If teeth are missing near the blade base and the leaf-tip is sharp, it is SWEET PEP-PER-BUSH.

2. If teeth are missing near the blade-base and the leaf-tip is broad, it is the CALIFORNIA BUCKTHORN.

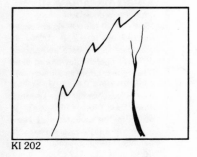

KI 202

3. If teeth occur near the blade-base, go to (KI 203), shown to the right, in the left-hand margin of page 405.

403

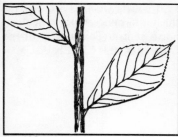

Fig. 402

SWEET PEPPER-BUSH
(Clethra alnifolia)
A shrub up to 9 ft. tall. Flowers tiny, white, borne in terminal clusters 3-8 in. long. The fruit is a rounded, dry capsule 1/8 in. in diameter. It grows in wet, acid forests and swamps in Louisiana and southeastern Texas. It belongs to the white alder family *(Clethraceae)*.

Fig. 403

CALIFORNIA BUCKTHORN
(Rhamnus californica)
Mostly a shrub, rarely a tree up to 18 ft. tall. Flower inconspicuous, greenish; fruit berry-like, green to red, ¼ in. in diameter. It grows in sheltered canyons in the mid-montane regions of New Mexico. It belongs to the buckthorn family *(Rhamnaceae)*.

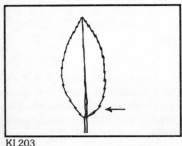

KI 203

404

1. If your specimen occurs as a thicket-forming shrub in central-north central Texas or southern Oklahoma, it is CREEK PLUM.

2. If your specimen occurs as a shrub and the leaf stalk is channeled, it is WINTERBERRY.

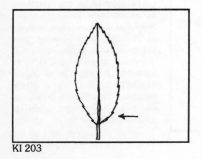

KI 203

3. If your specimen is a non-thicket forming shrub, the leaf stalk is unchanneled and it occurs in east Texas or Arkansas, it is LANCE-LEAVED BUCKTHORN.

4. If your specimen is a non-thicket forming shrub, with an unchanneled leaf stalk and you have found it in eastern Louisiana, it is STEWARTIA.

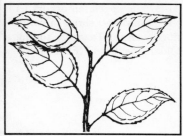

Fig. 404

Fig. 405

Fig. 406

Fig. 407

CREEK PLUM
(Prunus rivularis)
A shrub up to 6 ft. tall, forming dense thickets. The flower is white, ½ in. across and bearing many stamens. The fruit is red, cherry-like and ½ in. in diameter. It grows at forest edges, along creeks, roadsides or slope bottoms in central to north-central Texas and southern Oklahoma. It belongs to the rose family *(Rosaceae)*.

WINTERBERRY
(Ilex verticillata)
Shrub or small tree up to 25 ft. tall. Flower inconspicuous, greenish-white. Fruit lustrous, red to orange, ¼ in. in diameter, persisting through the winter. It grows in wet forests or swamps in Arkansas and southeastern Louisiana. It belongs to the holly family *(Aquifoliaceae)*.

LANCE-LEAVED BUCKTHORN
(Rhamnus lanceolata)
Shrub up to 9 ft. tall with greenish flowers ⅛ in. across. The fruit is a black olive ⅓ in. long. It grows in floodplain forests in eastern Texas and Arkansas. It belongs to the buckthorn family *(Rhamnaceae)*.

STEWARTIA
(Stewartia malacodendron)
A shrub or tree up to 18 ft. tall. Flower showy, white with many bluish to purple stamens, about 4 in. across. The fruit is a dry capsule ¾ in. in diameter. It grows in rich forests in eastern Louisiana. It is in the camellia family *(Theaceae)*.

1. If the leaves of your specimen are compound, go to (KI 205), shown to the right, in the left-hand margin of page 409.

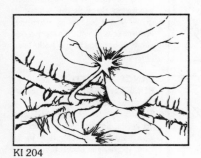

KI 204

2. If the leaves of your specimen are simple, find the symbol (KI 210), to the right, in the left-hand margin of page 419.

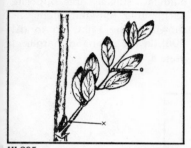

KI 205

KI 210

1. If the leaf stalks (not the bladelet stalks) are opposite each other on the stem, as shown to the right, go to (KI 206), in the left-hand margin of page 411.

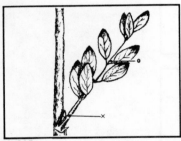

KI 205

2. If the leaf stalks (again, do not look at the bladelet stalks) are not opposite each other, find the picture to the right, (KI 207), in the left-hand margin of page 413. *DANGER:* Your specimen may be poisonous to the touch.

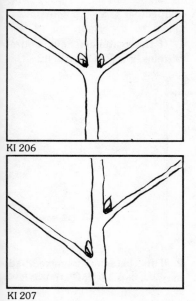

KI 206

KI 207

1. If the bladelets have entire margins, it is CROSS-VINE.

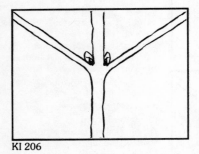

KI 206

2. If the bladelets have toothed margins, it is TRUMPET-VINE.

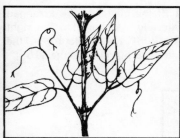

Fig. 408

Fig. 409

CROSS-VINE

(Bignonia capreolata)

A hairless, evergreen vine, climbing by tendrils. It grows up to 60 ft. long. Its flowers are reddish-orange, long trumpet-shaped with a yellow throat and about 2 in. long. The fruit is a 4-7 in. long capsule, opening to release uniquely winged seeds. It climbs in trees of bottomland forests of eastern Texas and Oklahoma pinelands, Louisiana and Arkansas. It belongs to the bignonia family *(Bignoniaceae)*.

TRUMPET-VINE

(Campsis radicans)

A vine climbing or not with aerial rootlets and up to 30 ft. long. Flower orange-red, tubular, 2-3½ in. long. The fruit is a dry capsule 2-6 in. long, bearing uniquely winged seeds. It grows in forests, over thickets and along fence rows in eastern Oklahoma and Texas east through Arkansas and Louisiana. It belongs to the bignonia family *(Bignoniaceae)*.

1. If the bladelets are arranged laterally, as shown to the right (KI 208), go to picture in the left-hand margin of **page** 415.

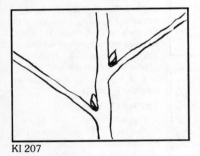

KI 207

2. If the bladelets are arranged from a point, as shown to the right (KI 209), go to this diagram in the left-hand margin of page 417.

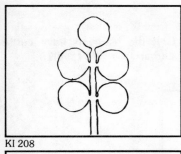

KI 208

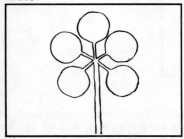

KI 209

1. If the bladelets have entire margins, it is WISTERIA.

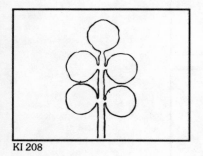

KI 208

2. If the bladelets are toothed, it is AMPELOPSIS.

Fig. 410

WISTERIA

(Wisteria frutescens var. macrostachya)

A thick-stemmed vine up to 40 ft. long. The lilac-purple pea flowers are ¾ in. long and borne in clusters 2-7 in. long. The fruit is a pod 1½-4 in. long. It grows in streamside forests in Arkansas, Louisiana and eastern Texas. It is in the pea family *(Leguminosae)*.

Fig. 411

AMPELOPSIS

(Ampelopsis arborea)

A smooth, red- to green-barked, slender vine, climbing by tendrils. The tiny, greenish-white flowers are inconspicuous. The fruit is a black berry ⅓ in. in diameter. It grows on moist soil in the eastern half of Texas including the panhandle, in Oklahoma, Arkansas and Louisiana. It belongs to the grape family *(Vitaceae)*.

1. If the leaflets are three, it is POISON IVY.

2. If the leaflets are seven, it is TEXAS WOODBINE.

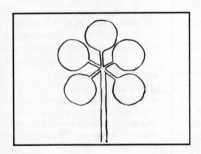

3. If the leaflets are five and none of the tendrils have disc (plate-like) tips, it is WESTERN WOODBINE.

4. If the leaflets are five and some of the tendrils have plate-like tips, it is VIRGINIA CREEPER.

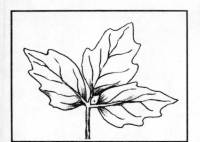

Fig. 412

POISON IVY
See page 57.

Fig. 413

TEXAS WOODBINE
(Parthenocissus heptaphylla)
A hairless, tendriled vine up to 30 ft. long. The tiny flowers are inconspicuous. The fruit is a bluish-black berry ½ in. in diameter. It is found in rocky or sandy soils, climbing on trees in the Edwards Plateau region of Texas. It belongs to the grape family *(Vitaceae)*.

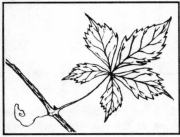

Fig. 414

WESTERN WOODBINE
(Parthenocissus vitacea)
A trailing or climbing vine. The tiny flowers are inconspicuous, becoming bluish-black berries 2/5 in. in diameter. It grows in streamside forests of western Texas and New Mexico. It belongs to the grape family *(Vitaceae)*.

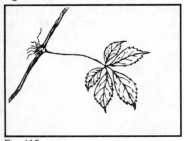

Fig. 415

VIRGINIA CREEPER
(Parthenocissus quinquefolia)
A high-climbing vine by means of disc-tipped tendrils. The tiny, greenish flowers are inconspicuous. The fruit is a bluish berry ⅓ in. in diameter. It is wide-ranging in the forests of Oklahoma, Arkansas, Louisiana and the eastern half of Texas. It belongs to the grape family *(Vitaceae)*.

418

1. If the leaf stalks *(not the bladelet stalks)* are opposite each other on the stem, as shown to the right in (KI 211), go to this drawing on page 421.

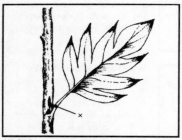

KI 210

2. If the leaf stalks (again do not look at the bladelet stalks) are not opposite each other, find the picture (KI 215), to the right, in the left-hand margin of page 429.

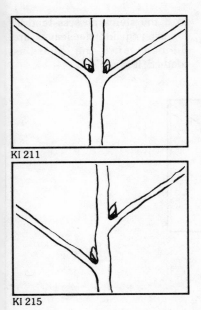

KI 211

KI 215

1. If the stems possess rootlet outgrowths for clinging and the leaf margins are entire, it is DECUMARIA VINE.

2. If the stems possess tendrils for clinging and the leaves are 3-lobed and the margins toothed, it is CISSUS.

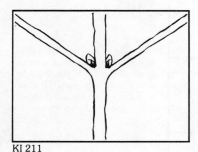

KI 211

3. If the rootlets or tendrils are lacking in your specimen, go to (KI 212), shown to the right, on page 423.

421

Fig. 416

DECUMARIA VINE
(Decumaria barbara)
A smooth, slender vine, bearing aerial rootlets. The small, fragrant, white flowers are borne in terminal clusters up to 4 in. wide. The fruit is a dry capsule ¼ in. long. It grows in swamps or wet, rich soil in Louisiana. It belongs to the saxifrage family *(Saxifragaceae)*.

Fig. 417

CISSUS
(Cissus incisa)
A warty-barked vine with simple tendrils and thick, fleshy leaves that have a disagreeable odor if crushed. The flower is greenish tiny, giving rise to black berries about ⅓ in. in diameter. It grows in upland forests throughout our area except New Mexico. It belongs to the grape family *(Vitaceae)*.

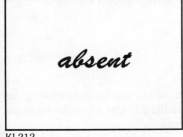

absent

KI 212

1. If the leaves are hairy on the undersurface and the upper leaves are joined by their bases, as shown to the right, it is GRAPE HONEYSUCKLE.

2. If the leaves are hairy on both surfaces and stalked, it is JAPANESE HONEYSUCKLE.

absent

KI 212

3. If the leaves are hairy only on their undersurface veins and stalked, it is TWINBERRY HONEYSUCKLE.

4. If the leaves are hairless, go to (KI 213), at the right, on page 425.

423

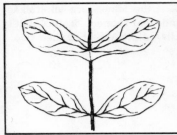
Fig. 418

Fig. 419

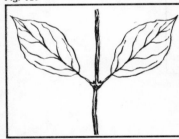
Fig. 420

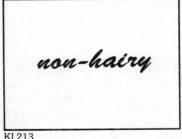
non-hairy

GRAPE HONEYSUCKLE
(Lonicera prolifera)
A twining vine 4-12 ft. long with whitened stems and leaves. Its flowers are showy, fragrant, tubular, yellow and 1 in. long. The fruit is a red berry ⅓ in. in diameter. It belongs to the honeysuckle family *(Caprifoliaceae)*.

JAPANESE HONEYSUCKLE
(Lonicera japonica)
A shreddy-barked, trailing or climbing vine up to 20 ft. long. Flowers fragrant, white to pink or becoming yellow, tubular with 2 flared lips, 1 in. long. The fruit is a black berry ½ in. in diameter. Introduced from Asia, this ornamental has escaped to become a troublesome weed in our area except for western Texas and New Mexico. It belongs to the honeysuckle family *(Caprifoliaceae)*.

TWINBERRY HONEYSUCKLE
(Lonicera involucrata)
A shrubby-twining vine forming thickets. Its yellow flowers are ½ in. long and subtended in pairs by 4 reddish floral leaves. The black fruit is a berry ⅓ in. in diameter. It grows in cold soil high in the mountains of Oklahoma and New Mexico. It is in the honeysuckle family *(Caprifoliaceae)*.

KI 213

1. If the leaves are all stalked and tear-shaped, it is CAROLINA JESSAMINE.

2. If the leaves are stalked and egg- to elliptic-shaped in outline, it is UTAH HONEYSUCKLE.

non-hairy

KI 213

3. If some of the leaves lack stalks or their bases are joined, go to (KI 214), shown to the right, in the left-hand margin of page 427.

425

Fig. 421

CAROLINA JESSAMINE
(Gelsemium sempervirens)
A twining, evergreen vine with slender, wiry stems. The fragrant, yellow, funnelform flowers are 1-1½ in. long. The oval capsule is flattened and ⅔ in. long, bearing winged seeds. It is found in moist, sandy soil in Arkansas, Oklahoma, Louisiana and eastern Texas. It belongs to the logania family *(Loganiaceae)*.

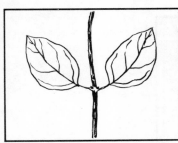

Fig. 422

UTAH HONEYSUCKLE
(Lonicera utahensis)
A twining shrub up to 5 ft. tall, forming colonies. The yellow, tubular flower is sometimes purple-tinged, fragrant and about 1 in. long. The berry is orange to red and ¼ in. in diameter. It grows in the pine forests of the mountains of New Mexico. It belongs to the honeysuckle family *(Caprifoliaceae)*.

KI 214

1. If the upper base-joined leaves have notched tips, as shown to the right, it is YELLOW HONEYSUCKLE.

2. If none of the leaf-tips are notched but the base-joined pairs are incompletely joined as shown to the right, it is WESTERN HONEYSUCKLE.

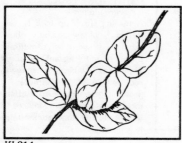

KI 214

3. If none of the leaf tips are notched but the base-joined pairs are completely joined and your specimen occurs in east Texas, Oklahoma, Arkansas or Louisiana, it is TRUMPET HONEYSUCKLE.

4. If none of the leaf tips are notched but the base-joined pairs are completely joined and your specimen occurs in west Texas or New Mexico, it is ARIZONA HONEYSUCKLE.

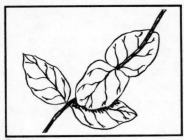

Fig. 423

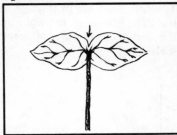

Fig. 424

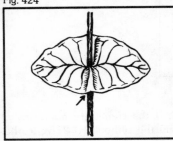

Fig. 425

Fig. 426

YELLOW HONEYSUCKLE
(Lonicera flava)

A hairless, climbing or trailing vine. The conspicuous flowers are cylindric, bright orange-yellow, ¾-1¼ in. long and 2-lipped. The fruit is a reddish berry ¼ in. in diameter. It grows in rocky, upland forests in Oklahoma and Arkansas. It belongs to the honeysuckle family *(Caprifoliaceae)*.

WESTERN HONEYSUCKLE
(Lonicera albiflora)

A bushy plant with twining branches up to 9 ft. tall. Its 2-lipped flowers are slender, white, and ⅔ in. long. The fruit is a hairy berry 2/5 in. in diameter. It grows on rocky and sandy soils in north-central to central Texas and Oklahoma. It belongs to the honeysuckle family *(Caprifoliaceae)*.

TRUMPET HONEYSUCKLE
(Lonicera sempervirens)

A twining, evergreen shrubby vine up to 18 ft. long. Its flowers are slender, trumpet-shaped, red or yellow, 1½-2 in. long. The fruit is a small, red berry. It grows in upland forests in Oklahoma, Arkansas, Louisiana and eastern Texas. It belongs to the honeysuckle family *(Caprifoliaceae)*.

ARIZONA HONEYSUCKLE
(Lonicera arizonica)

A stiff climbing or trailing vine with ¾-1½ in. long, red, short-lobed flowers. The fruit is a small, red berry. It grows in the coniferous forests of far western Texas and New Mexico. It belongs to the honeysuckle family *(Caprifoliaceae)*.

428

1. If the stems are prickly, go to (KI 216), shown at the right, in the left-hand margin of page 431.

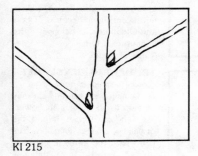

KI 215

2. If the stems are non-prickly, find the symbol (KI 218), to the right, in the left-hand margin of page 435.

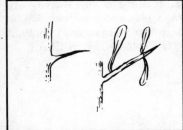

KI 216

KI 218

1. If the leaves are triangular in shape with some prickles on their margins, it is SAW GREENBRIAR.

2. If some of the leaves are triangular and some egg-shaped and their margins are smooth, it is COMMON GREENBRIAR.

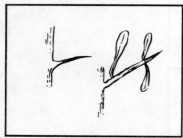

KI 216

3. If the leaves are broad with parallel margins, it is LAUREL GREENBRIAR.

4. If the leaves are egg- to elliptic-shaped in outline, go to (KI 217), shown to the right, in the left-hand margin of page 433.

431

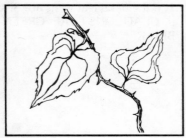
Fig. 427

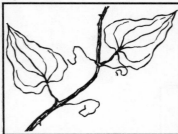
Fig. 428

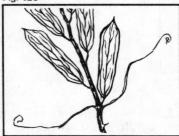
Fig. 429

KI 217

SAW GREENBRIAR
(Smilax bona-nox)
A rampant climbing or trailing vine with 4-angled stems beset with recurved prickles. Flowers tiny, greenish, imperfect. The fruit is a black berry ¼ in. in diameter. It grows in upland forests or cleared old fields and roadsides throughout our area except for far western Texas and New Mexico. It belongs to the lily family *(Liliaceae)*.

COMMON GREENBRIAR
(Smilax rotundifolia)
An evergreen, high-climbing vine with round to angled prickled stems and numerous tendrils. The tiny, greenish-yellow flowers are imperfect. The fruit is a black berry about ¼ in. in diameter borne in tight clusters of 5-20 berries. It is widespread in forests, fence rows and thickets, often forming noxious tangles, in Oklahoma, Arkansas, Louisiana and most of Texas except for the panhandle and far western areas. It belongs to the lily family *(Liliaceae)*.

LAUREL GREENBRIAR
(Smilax laurifolia)
An evergreen, rampant, high-climbing vine with round stems, thick, stout spines and paired tendrils. The greenish-yellow flowers are imperfect and very small. The fruits are black berries, ⅓ in. in diameter borne in large, loose clusters. It grows on acid sands in humid forests or swamps in Louisiana, Arkansas and eastern Texas. It is in the lily family *(Liliaceae)*.

1. If the basal leaf veins are 3, it is GLAUCOUS-LEAF GREEN-BRIAR.

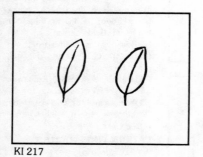

KI 217

2. If the leaf veins are 5 to 7, it is the DEVIL'S GREENBRIAR.

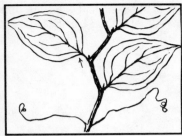

Fig. 430

GLAUCOUS-LEAF GREENBRIAR

(Smilax glauca)

A tangled, free-climbing vine with round stems bearing abundant spines and delicate tendrils. The very small, greenish flowers are not too conspicuous. The fruit is a roundish cluster of black berries, each ⅓ in. in diameter. It grows on dry sandy soil in Oklahoma, Arkansas, Louisiana and eastern Texas. It belongs to the lily family *(Liliaceae)*.

Fig. 431

DEVIL'S GREENBRIAR

(Smilax hispida)

A stout, high-climbing vine with the lower stems covered with needle-like spines. The small flowers are imperfect and greenish. The fruit is a black berry ⅓ in. in diameter. It grows in streamside and swamp forests in Oklahoma, Arkansas, Louisiana, north-central and eastern Texas. It belongs to the lily family *(Liliaceae)*.

434

1. If your specimen lacks ten-drils, go to (KI 219), shown to the right, in the left-hand margin of pages 437 and 437A.

KI 218

2. If tendrils are present in your specimen, look for the diagram (KI 220), shown to the right, in the left-hand margin of page 439.

absent

KI 219

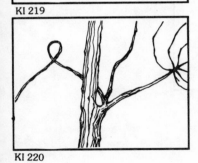

KI 220

1. If the leaves are elliptic- to egg-shaped, hairless and the margins smooth, it is SUPPLEJACK.

2. If the leaves are elliptic-shaped, hairless, and most of the margins are toothed, it is CLIMBING BITTERSWEET.

absent

KI 219

3. If some of the leaves are heart-shaped, hairless, and have notched tips, it is DIVERSE-LEAF MOONSEED.

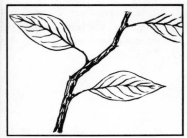
Fig. 432

SUPPLEJACK
(Berchemia scandens)
A high-climbing, pliant vine. The small flowers are yellowish-green and the fruit is a black olive (drupe) ⅓ in. long. It grows in moist soils in forests of Oklahoma, Arkansas, Louisiana and the eastern half of Texas. It belongs to the buckthorn family *(Rhamnaceae)*.

Fig. 433

CLIMBING BITTERSWEET
(Celastrus scandens)
A high-climbing vine up to 45 ft. long. The flowers are small and greenish. The fruit is an orange capsule 2/5 in. in diameter, splitting to reveal red seeds. It grows in canyons, upland and streamside forests, thickets and fence rows in Texas, Oklahoma, Arkansas and Louisiana. It belongs to the bittersweet family *(Celastraceae)*.

Fig. 434

DIVERSE-LEAF MOONSEED
(Cocculus diversifolius)
A slender, short vine up to 6 ft. long. The small flowers are greenish-white, producing a dark purple, fleshy fruit ¼ in. in diameter, bearing a single seed coiled like a snail shell. It grows in brushland in southern and western Texas and the mountains of southern New Mexico. It belongs to the moonseed family *(Menispermaceae)*.

438

4. If the leaves are heart-shaped and quite hairy, it is the PIPEVINE.

absent

KI 219

5. If the leaves are diamond-shaped, it is MAGNOLIA VINE.

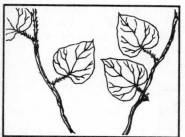

Fig. 435

PIPEVINE

(Aristolochia tomentosa)
High-climbing vine up to 75 ft.
long. Flowers pipe-shaped, tube
yellowish-green, petals dark purple, length 1-2 in. The fruit is an
oblong capsule 4 in. long. It
grows in rich forests in Louisiana,
Arkansas, Oklahoma and eastern
Texas. It belongs to the birthwort
family *(Aristolochiaceae)*.

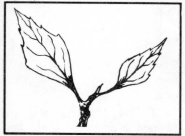

Fig. 436

MAGNOLIA VINE

(Schizandra coccinea)
A slender vine with crimson-
purple, imperfect flowers ¼-½
in. across. The fruit is orange-red
and about 1/5 in. in diameter. It
grows in rich forests in Arkansas
and Louisiana. It is in the
magnolia family *(Magnoliaceae)*.

1. If the leaf margins are entire, go to (KI 221), shown to the right, in the left-hand margin of page 441.

KI 220

2. If the leaf margins are toothed or lobed or both, find the drawing to the right (KI 222), in the left-hand margin of page 443.

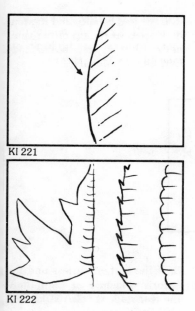

KI 221

KI 222

1. If the leaf is hairy and the tendrils are attached to the stem at the leaf base, as shown to the right, it is DWARF GREENBRIAR.

2. If the leaf is hairy and the tendrils occur at the tip of the vine, never at the leaf base, it is CAROLINA MOONSEED.

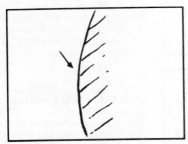

KI 221

3. If the leaf is hairless and the tendrils originate on the stem at the leaf stalk, it is LANCELEAF GREENBRIAR.

4. If the leaf is hairless, and the tendrils originate at the end of a branch, it is BUCKWHEAT VINE.

441

Fig. 437

DWARF GREENBRIAR
(Smilax pumila)
A slender, weak, mostly non-climbing vine with tiny, inconspicuous flowers. The fruit is fleshy, red to orange, shiny and ⅓ in. in diameter. It grows in pine and oak forests and along streams in Arkansas, Louisiana and eastern Texas. It belongs to the lily family *(Liliaceae)*.

Fig. 438

CAROLINA MOONSEED
(Cocculus carolinus)
A slender vine with small, greenish-white flowers and clusters of conspicuous red, fleshy fruits, each about ¼ in. long and bearing a pit curled into a spiral like a snail shell. This short vine (up to 9 ft. long) is common in Oklahoma, Arkansas, Louisiana and the eastern half of Texas. It belongs to the moonseed family *(Menispermaceae)*.

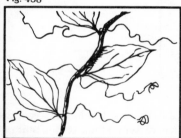

Fig. 439

LANCELEAF GREENBRIAR
(Smilax lanceolata)
A stout, evergreen vine to 40 ft. long. The tiny, imperfect flowers are yellowish-green. The fruit is a brownish berry ¼ in. in diameter. It grows in moist forests and fields in Arkansas, Louisiana and eastern Texas. It belongs to the lily family *(Liliaceae)*.

Fig. 440

BUCKWHEAT VINE
(Brunnichia ovata)
A vine up to 40 ft. long with tubular, greenish flowers ¾-1½ in. long borne in terminal clusters up to 10 in. long. The fruit is tiny, seed-like and triangular in shape. It is infrequent in moist, rich forests in Oklahoma, Arkansas, Louisiana and eastern Texas. It belongs to the buckwheat family *(Polygonaceae)*.

442

1. If the leaf is fleshy and 3-parted to the point it almost appears compound, as shown to the right, it is CISSUS.

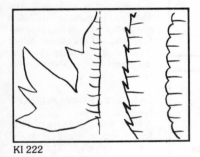

KI 222

2. If the leaf is thin and shallow-ly, 3-5 lobed (mostly 3-lobed) or rarely unlobed, it is one of our some 13 kinds of grapes, so go to (KI 223), shown to the right, in the left-hand margin of page 445.

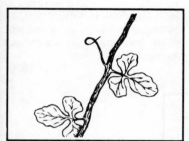

Fig. 441

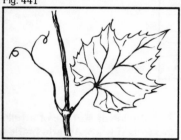

KI 223

CISSUS

(Cissus incisa)

A warty-barked vine with simple tendrils and thick, fleshy leaves that have a disagreeable odor if crushed. The tiny flower is greenish, giving rise to black berries about ⅓ in. in diameter. It grows in upland forests throughout our area except for New Mexico. It belongs to the grape family *(Vitaceae)*.

1. If the lower surface of a fully grown leaf is covered with a continuous or patchy mass of cobwebby hairs, go to (KI 224), as shown to the right, in the left-hand margin of pages 447 and 447A.

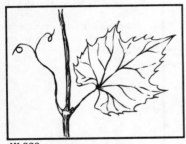

KI 223

2. If the lower surface of a fully-grown leaf is apparently hairless (not counting the ribs), go to the symbol (KI 225), shown to the right, in the left-hand margin of page 449.

KI 224

non-hairy

KI 225

1. If the undersurface mat of hairs is snow-white and the leaves are barely lobed, as shown to the right, it is the MUSTANG GRAPE.

2. If the undersurface mat of hairs is a bluish-ashen color and the leaf lobes are evident, as shown to the right, it is the PANHANDLE GRAPE.

KI 224

3. If the undersurface mat of hairs is gray and the leaves are scarcely lobed, as shown to the right, it is the GRAYBARK GRAPE.

Fig. 442

MUSTANG GRAPE
(Vitis mustangensis)
A high-climbing vine 40 ft. or longer. Flower tiny, white becoming a purple-black grape ½-4/5 in. in diameter. It grows along stream bottoms, thickets, fence rows and forest edges in western Louisiana, Arkansas, Oklahoma and the eastern half of Texas. It is in the grape family *(Vitaceae)*.

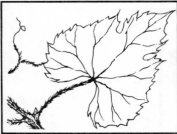

Fig. 443

PANHANDLE GRAPE
(Vitis acerifolia)
Stocky vine, rarely climbing, but often covering shrubs or rock piles. Flower white, inconspicuous. The grape is black, about ½ in. in diameter and covered with a white bloom. It grows on rocky slopes or in open woods in western Oklahoma, through the Texas panhandle and into eastern New Mexico. It belongs to the grape family *(Vitaceae)*.

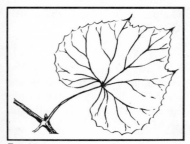

Fig. 444

GRAYBARK GRAPE
(Vitis cinerea)
A lax, high-climbing, large vine. The tiny, white flowers are inconspicuous. The grape is black to purple and up to ⅓ in. in diameter. It grows in rich bottomland forests in Oklahoma, Arkansas, Louisiana and north-central to eastern Texas. It belongs to the grape family *(Vitaceae)*.

448

4. If the leaf undersurface hairs are rusty-colored and your specimen occurs near a stream, it is the SUMMER GRAPE.

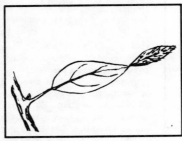

KI 224

5. If the leaf undersurface hairs are rusty-colored and your specimen occurs in upland woods, it is the POST OAK GRAPE.

447A

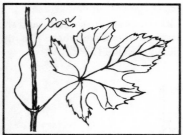

Fig. 445

SUMMER GRAPE
(Vitis aestivalis)
A vigorous, high-climbing vine
with inconspicuous white flowers.
The grape is dark blue to black
and up to ½ in. in diameter. It
grows in dry, sandy forests,
thickets or along roadsides in
Oklahoma, Arkansas, Louisiana
and the eastern third of Texas. It
is in the grape family *(Vitaceae)*.

Fig. 446

POST OAK GRAPE
(Vitis lincecomii)
A moderate climbing vine in
small trees or forming a bushy
clump in the open. The flowers
are inconspicuous and the grape
is black and 2/5-1 in. in
diameter. It grows in open, sandy
forests in east and south-central
Texas, Louisiana, Oklahoma and
Arkansas. It is in the grape family *(Vitaceae)*.

448A

1. If the bark is tight (never shreddy), it is the MUSCADINE GRAPE.

2. If the bark is shreddy and the leaves are sharp-lobed, it is the CATBIRD GRAPE.

non-hairy

KI 225

3. If the leaves are unlobed or shallowly lobed, go to (KI 226), on the right, in the left-hand margin of page 451.

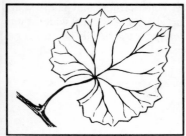

Fig. 447

MUSCADINE GRAPE
(Vitis rotundifolia)
Large, vigorous vines up to 100 ft. long, lacking the shreddy bark of most grapes. Flowers inconspicuous, the grape black, ⅓ in. in diameter. It grows in open sandy forests in Oklahoma, Arkansas, Louisiana, eastern and south-central Texas. It belongs to the grape family *(Vitaceae)*.

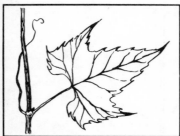

Fig. 448

CATBIRD GRAPE
(Vitis palmata)
A high-climbing, slender vine with tiny, inconspicuous, white flowers and grapes that are bluish-black, ¼ in. in diameter. It grows in sloughs, pond margins and rich, low forests in Louisiana, Arkansas, eastern Oklahoma and eastern Texas. It belongs to the grape family *(Vitaceae)*.

KI 226

1. If the largest leaves are 3 in. or shorter and are wider than long, it is the SAND GRAPE.

2. If the largest leaves are 3 in. or shorter and are as long as wide, it is the SWEET MOUNTAIN GRAPE.

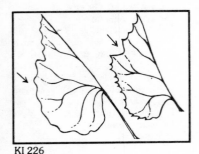

KI 226

3. If the largest leaves are more than 3 inches long, go to (KI 227), shown to the right, in the left-hand margin of page 453.

451

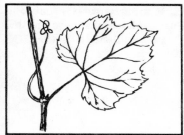

Fig. 449

SAND GRAPE
(Vitis rupestris)
A mostly non-climbing vine, forming a tangle over rock piles or low shrubs. Flowers white, inconspicuous with black grapes ¼-½ in. in diameter. It grows in sand hills or in limestone soils in central and southwestern Texas into Oklahoma and across into Arkansas. It belongs to the grape family *(Vitaceae)*.

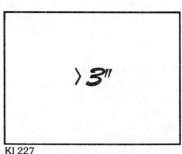

Fig. 450

SWEET MOUNTAIN GRAPE
(Vitis monticola)
A slender, climbing vine to 30 ft long. Flowers white, inconspicuous, becoming black grapes ⅛-½ in. in diameter. It grows on limestone hills and ridges in the Edward's Plateau region of Texas. It is in the grape family *(Vitaceae)*.

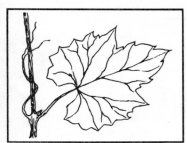

KI 227

452

1. If the undersurface leaf veins are hairless or if hairy and the hairs are short, it is the FOX GRAPE.

2. If the undersurface leaf veins are clothed with long hairs and your specimen occurs on the Edwards Plateau in Texas, it is the WINTER GRAPE.

KI 227

3. If the undersurface leaf veins are clothed with long hairs and your specimen occurs in east Texas, Oklahoma, Arkansas or Louisiana, it is the FROST GRAPE.

4. If the undersurface leaf veins are clothed with long hairs and your specimen occurs in the mountains of Trans-Pecos Texas and New Mexico, it is the CANYON GRAPE.

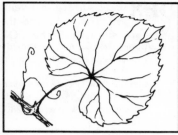

Fig. 451

FOX GRAPE
(Vitis vulpina)

A vigorous, large-trunked, high-climbing vine with tiny, inconspicuous, white flowers and black, shiny grapes, ¼ in. in diameter. It grows in ravines and canyons of the Chios and Black Mountains of southwestern Texas and western New Mexico. It belongs to the grape family *(Vitaceae)*.

Fig. 452

WINTER GRAPE
(Vitis berlandieri)

Moderately climbing vine with inconspicuous, white flowers and black to purple grapes about ⅓ in. in diameter. It grows in limestone or sandy soils in central and southwestern Texas. It is in the grape family *(Vitaceae)*.

Fig. 453

FROST GRAPE
(Vitis cordifolia)

A large, high-climbing vine, often overtopping the canopy of tall trees. Its flowers are tiny and white. Its dull black grapes are 1/8-2/5 in. in diameter. It grows in floodplain forests in central and eastern Oklahoma, central and eastern Texas, Louisiana and Arkansas. It belongs to the grape family *(Vitaceae)*.

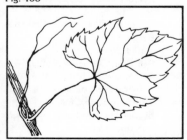

Fig. 454

CANYON GRAPE
(Vitis arizonica)

A grayish, small, weak vine with tiny, white, inconspicuous flowers. The black grape is 2/5 in. in diameter. It grows in ravines in far western Texas and New Mexico. It belongs to the grape family *(Viyaceae)*.

454

INDEX

Acacia, *Acacia spp.*, 70, 72
  Prairie, 104
Adolphia
  Texas, 6
Agave, *Agave spp.*, 110
Alder
  Mexican, 300, 374
  Oklahoma, 296, 388
  Smooth, 334
Alder-leaved Service-berry,
    Amelanchier alnifolia, 332
Allspice
  Carolina, 174
Aloysia, *Aloysia wrightii*, 136
Alpine Fir, *Abies lasiocarpa*, 34
American Basswood, *Tilia americana*,
    290
American Bladdernut, *Staphylea
    trifolia*, 50
American Elm, *Ulmus americana*,
    288
American Holly, *Ilex opaca*, 322
American Hornbeam, *Carpinus
    caroliniana*, 372
American Plum, *Prunus americana*,
    338
Ampelopsis, *Ampelopsis arborea*,
    416
Anacua, *Ehretia anacua*, 216
Andrachne
  Northern, 232
Anisacanth, *Anisacanthus wrightii*,
    178
Apache Plume, *Fallugia paradoxa*,
    280
Arizona Cypress, *Cupressus
    arizonica*, 16
Arizona Honeysuckle, *Lonicera
    arizonica*, 428
Arizona Madrone, *Arbutus arizonica*,
    202

Arizona Oak, *Quercus arizonica*, 346
Arizona Plane-tree, *Platanus wrightii*,
    364
Arkansas Oak, *Quercus arkansana*,
    242
Arrow-weed, *Tessaria sericea*, 230
Arrowwood
  Southern, 138
Ash, *Fraxinus spp.*, 44
  Mountain, 84
  Prickly, 76
Aspen
  Quaking, 296
Autumn Sage, *Salvia greggii*, 166
Azalea, *Rhododendron spp.*, 346

Baccharis
  Sticky, 320
Bald Cypress, *Taxodium distichum*,
    32
Bamboo
  Cane, 278
Basswood
  American, 290
  Carolina, 290A
Bay
  Red, 240
Beach Croton, *Croton punctatus*, 256
Beachwort, *Batis maritima*, 154
Beaked Willow, *Salix bebbiana*, 210
Bearberry, *Arctostaphylos uva-ursi*,
    242
Bee-brush, *Aloysia gratissima*, 166
Beech
  Red, 384
Bee Tree, *Tilia heterophylla*, 290
Bernardia
  Johnston's, 282
  Southwestern, 282
Big Bend Service-berry, *Amelanchier
    denticulata*, 386

Big Saltbush, *Atriplex lentiformis,* 254

Big-Tooth Maple, *Acer grandidentatum,* 122

Birch, *Betula nigra,* 364

Birch-leaf Buckthorn, *Rhamnus betulaefolia,* 376

Bitter Brush, *Purshia tridentata,* 362

Bittersweet
Climbing, 438

Black-bush, *Coleogyne ramosissima,* 114

Black Cherry, *Prunus serotina,* 400

Black-gum, *Nyssa sylvatica,* 274

Blackhaw, *Viburnum prunifolium,* 130

Black Huckleberry, *Gaylussacia baccata,* 248, 268

Blackjack Oak, *Quercus marilandica,* 342

Black Locust, *Robinia pseudo-acadia,* 74A

Black-mangrove, *Avicennia germinans,* 164

Black Maple, *Acer nigrum,* 120

Black Oak, *Quercus velutina,* 312, 392

Black Walnut, *Juglans nigra,* 86

Black Willow, *Salix nigra,* 356

Bladdernut
American, 50

Blood-of-christ, *Jatropha cardiophylla,* 318

Blueberry, *Vaccinium amoenum* and *virgatum,* 282
Blueridge, 196
Dwarf, 258
Elliott, 196
Evergreen, 244
Thick-leaved, 270

Blueridge Blueberry, *Vaccinium vacillans,* 196

Blue Sage, *Salvia ballotoeflora,* 140

Bluet

Shrub, 158

Bluestem Willow, *Salix irrorata,* 202

Bluewood, *Condalia hookeri,* 190A
Mexican, 184

Bois D'arc, *Maclura pomifera,* 190

Bouchea
Flaxleaf, 160

Bouvardia
Scarlet, 156

Boxelder, *Acer negundo,* 44

Bramble, *Rubus spp.,* 74

Brickellia, *Brickellia squamulosa,* 168

Bristlecone Pine, *Pinus aristata,* 24

Buck-brush, *Ceanothus fendleri,* 190

Buckeye, *Aesculus spp.,* 50
Mexican, 84A

Buckthorn, *Rhamnus spp.,* 304
Birch-leaf, 376
California, 208A, 404
Green, 190A
Lance-leaved, 406
Texas, 186A
Wooly, 114

Buckwheat Vine, *Brunnichia ovata,* 442

Buffalo-berry
Canadian, 174
Silver, 114

Bur Oak, *Quercus macrocarpa,* 350

Bush Huckleberry, *Gaylussacia dumosa,* 230A

Butterfly-bush, *Buddleja scordioides,* 132

Butternut, *Juglans cinerea,* 86

Buttonbush, *Cephalanthus occidentalis,* 174

Cactus
Deerhorn, 8

California Buckthorn, *Rhamnus californica,* 208A, 404

California Redbud, *Cercis occidentalis,* 236

Canadian Bufflo-berry, *Sheperdia*

canadensis, 174
Candleberry, *Myrica cerifera*, 248
Cane Bamboo, *Arundinaria gigantea*, 278
Canyon Grape, *Vitis arizonica*, 454
Carolina Allspice, *Calycanthus floridus*, 174
Carolina Basswood, *Tilia caroliniana*, 290
Carolina Cherry-laurel, *Prunus caroliniana*, 202
Carolina Jessamine, *Gelsemium sempervirens*, 426
Carolina Moonseed, *Cocculus carolinus*, 442
Carolina Willow, *Salix caroliniana*, 300
Catalpa, *Catalpa speciosa*, 172
Catbird Grape, *Vitis palmata*, 450
Cedar Elm, *Ulmus crassifolia*, 374
Cenizo, *Leucophyllum frutescens*, 156
Century-plant, *Agave spp.*, 110
Chaguira, *Ceanothus integerrimus*, 246
Chaste-tree, *Vitex agnus-castus*, 48
Cherry
    Black, 400
Cherry-laurel
    Carolina, 202
Chestnut Oak, *Quercus prinus*, 344
Chinaberry, *Melia azedarch*, 84A
Chinquapin, *Castanea spp.*, 398
Chinquapin Oak, *Quercus muhlenbergii* and/or *prinoides*, 360
Chokecherry, *Prunus virginiana*, 338
Cholla, *Opuntia spp.*, 8
Cinquefoil
    Shrubby, 90
Cissus, *Cissus incisa*, 422, 444
Cliff Fendler-bush, *Fendlera rupicola*, 160
Cliff Jamesia, *Jamesia americana*, 138

Climbing Bittersweet, *Celastrus scandens*, 438
Coastal Leucothoe, *Leucothoe axillaris*, 304, 388
Coldenia, *Coldenia greggii*, 256
Colorado Blue Spruce, *Picea pungens*, 32
Common Greenbriar, *Smilax rotundifolia*, 432
Common Ninebark, *Physocarpus opulifolius*, 310
Coralberry, *Symphoricarpos spp.*, 178
Corkwood, *Leitneria floridana*, 258
Cortes Croton, *Croton cortesianus*, 266
Cottonwood, *Populus spp.*, 296
    Narrow-leaved, 320
    Swamp, 384
Coyote Willow, *Salix argophylla*, 280A
Coyotillo, *Karwinskia humboldtiana*, 176
Crabapple
    Southern, 186
    Western, 186
Creek Plum, *Prunus rivularis*, 406
Creosote Bush, *Larrea tridentata*, 172
Cross-vine, *Bignonia capreolata*, 412
Croton
    Beach, 256
    Cortes, 266
    Fragrant, 254
    Torrey, 258
Crucifixion Thorn, *Koeberlinia spinosa*, 6
Crucilla, *Randia rhagocarpa*, 114
Crucillo, *Condalia obtusifolia* var. *lycioides*, 184
Currant, *Ribes spp.*, 148
Cypress
    Arizona, 16
    Bald, 32
Cyrilla, *Cyrilla racemiflora*, 242A

Dahoon Holly, *Ilex cassine*, 242A

Dalea, *Dalea spp.*, 90

Damiana, *Turnera diffusa* var. *aphrodisiaca*, 280

Dangleberry, *Gaylussacia frondosa*, 266A

Deciduous Swamp Holly, *Ilex decidua*, 316

Decumaria Vine, *Decumaria barbara*, 422

Deerhorn Cactus, *Cereus greggii*, 8

Desert Honeysuckle, *Anisacanthus thurberi*, 160

Desert Olive, *Forestiera pubescens* var. *glabrifolia*, 128

Desert Peach, *Prunus fasiculata*, 184

Desert-willow, *Chilopsis linearis*, 144

Desert Youpon, *Schaefferia cuneifolia*, 232

Devil's Elbow, *Forestiera pubescens*, 132

Devil's Greenbriar, *Smilax hispida*, 434

Devil's Walking Stick, *Aralia spinosa*, 76

Diamond-leaf Oak, *Quercus laurifolia* var. *obtusa*, 236

Diverse-leaf Moonseed, *Cocculus diversifolius*, 438

Dogberry, *Pyrus arbutifolia*, 382

Dogwood
  Flowering, 172

Douglas Fir, *Pseudotsuga menziesii*, 32

Dune Broom, *Parryella filifolia*, 102

Durand Oak, *Quercus sinuata*, 212

Dwarf Blueberry, *Vaccinium depressum*, 258

Dwarf Greenbriar, *Smilax pumila*, 442

Dwarf Huckleberry, *Gaylussacia hirtella*, 252

Dwarf Palm, *Sabal minor*, 56

Dwarf Willow, *Salix tristis*, 252

Eastern Erythria, *Erythrina herbacea*, 74

Eastern Wahoo, *Euonymus atropurpureus*, 140

Ebony
  Texas, 70

Elderberry, *Sambucus spp.*, 44

Elliott Blueberry, *Vaccinium elliottii*, 196

Elm
  American, 288
  Cedar, 374
  Rock, 376
  September, 288
  Slippery, 288
  Water, 304
  Winged, 372

Emory Oak, *Quercus emoryi*, 354

Englemann Spruce, *Picea engelmannii*, 32

Ephedra, *Ephedra spp.*, 10

Erythrina, *Erythrina spp.*, 74
  Eastern, 74
  Western, 74

Euphorbia
  Wax, 10

Evergreen Blueberry, *Vaccinium darrowii*, 244

Evergreen Sumac, *Rhus virens*, 90

Evergreen Swamp Holly, *Ilex vomitoria*, 326

Eve's Necklace, *Sophora affinis*, 98

False Mesquite, *Calliandra conferta* or *eriophylla*, 100

Fendler-bush
  Cliff, 160

Fetterbush, *Lyonia lucida*, 238

Fiddlewood
  Mexican, 156

Fir
  Alpine, 34
  Douglas, 32
  White, 34

Flaxleaf St. Peter's-wort, *Ascyrum hypericoides*, 160A

Flaxleaf Bouchea, *Bouchea linifolia*, 160

Florida Linden, *Tilia floridana*, 290

Florida Sugar Maple, *Acer barbatum*, 120

Flowering Dogwood, *Cornus florida*, 172

Fox Grape, *Vitis vulpina*, 454

Fragrant Croton, *Croton suaveolens*, 254

Fragrant Sumac, *Rhus aromatica*, 58A

Frankenia, *Frankenia jamesii*, 158

French-mulberry, *Callicarpa americana*, 134

Fringe-tree, *Chionanthus virginica*, 176

Frost Grape, *Vitis cordifolia*, 454

Gambel Oak, *Quercus gambelii*, 348

Georgia Swamp Holly, *Ilex longipes*, 302

Glaucous-leaf Greenbriar, *Smilax glauca*, 434

Goat-bush, *Castela texana*, 150

Golden-ball Lead-tree, *Leucaena retusa*, 98

Goldeneye, *Viguiera deltoidea* var. *parishii*, 146

Grape
  Canyon, 454
  Catbird, 450
  Fox, 454
  Frost, 454
  Graybark, 448
  Muscadine, 450
  Mustang, 448
  Panhandle, 448
  Post Oak, 448A
  Sand, 452
  Summer, 448A
  Sweet Mountain, 452
  Winter, 454

Grape Honeysuckle, *Lonicera prolifera*, 424

Graybark Grape, *Vitis cinerea*, 448

Gray-leaf Willow, *Salix glauca*, 272

Gray Oak, *Quercus grisea*, 216

Grease-bush, *Forsellesia spinescens*, 192

Greaseleaf Sage, *Salvia pinguifolia*, 134

Greasewood, *Sarcobatus vermiculatus*, 192

Greenbriar
  Common, 432
  Devil's, 434
  Dwarf, 442
  Glaucous-leaf, 434
  Lanceleaf, 442
  Laurel, 432
  Saw, 432

Green Buckthorn, *Condalia viridis*, 190A

Gum
  Black, 274

Hackberry, *Celtis occidentalis*, 318, 402
  Sugar, 236A
  Thin-leaf, 264

Hairy Mockorange, *Philadelphus hirsutus*, 138

Haw
  Black, 130
  Possum, 168

Hawthorn
  Lobed, 186
  Smooth, 186A

Hazelnut, *Corylus americana*, 376

Heart-leaf Willow, *Salix cordata*, 356

Hickory
  Shagbark, 84
  Shellbark, 84A

Holly
  American, 322
  Dahoon, 242A
  Deciduous Swamp, 316
  Evergreen Swamp, 326
  Georgia Swamp, 302

459

Inkberry, 206
Sand, 326
Honey Locust, *Gleditsia tricanthos*, 64
Honey Mesquite, *Prosopis glandulosa*, 66
Honeysuckle
Arizona, 428
Desert, 160
Grape, 424
Japanese, 424
Trumpet, 428
Twinberry, 424
Utah, 426
Western, 428
Yellow, 428
Hop Hornbeam, *Ostrya knowltonii*, 388
Hop-tree, *Ptelea trifoliata*, 50, 58A
Hornbeam
American, 372
Hop, 388
Wooly, 376
Hortulan Plum, *Prunus hortulana*, 334
Huckleberry, *Gaylussacia* and *Vaccinium* spp., 274
Black, 248, 268
Bush, 230A
Dwarf, 252
Indigo-bush, *Amorpha spp.*, 94, 96, 98
Leaden, 104
Smooth, 98
Inkberry Holly, *Ilex glabra*, 206

Jamesia
Cliff, 138
Japanese Honeysuckle, *Lonicera japonica*, 424
Javelina Bush, *Condalia ericoides*, 148
Jerusalem Thorn, *Parkinsonia aculeata*, 70
Jessamine

Carolina, 426
Johnston's Bernardia, *Bernardia obovata*, 282
Juniper, *Juniperus spp.*, 16

Kidneywood, *Eysenhardtia polystachya*, 104
Texas, 100
Krameria, *Krameria spp.*, 14

Lacey Oak, *Quercus glaucoides*, 308
Lanceleaf Greenbriar, *Smilax lanceolata*, 442
Lance-leaved Buckthorn, *Rhamnus lanceolata*, 406
Lantana, *Lantana spp.*, 140
Large-leaf Magnolia, *Magnolia macrophylla*, 262
Laurel
Mountain, 244, 264
Laurel Greenbriar, *Smilax laurifolia*, 432
Laurel Oak, *Quercus laurifolia*, 208
Leaden indigo-bush, *Amorpha canescens*, 104
Lead-tree
Golden-ball, 98
Leatherwood, *Dirca palustris*, 230
Leucothae, *Leucothoe elongata* and *racemosa*, 382
Coastal, 304, 388
Limber Pine, *Pinus strobiformis*, 24
Limestone Oak, *Quercus mohriana*, 218
Linden
Florida, 290
Little-leaf Sumac, *Rhus microphylla*, 90A
Live Oak, *Quercus virginiana*, 208A, 222
Lobed-hawthorn, *Crataegus spp.*, 186
Loblolly Pine, *Pinus taeda*, 30
Locust
Honey, 64

New Mexican, 74A
Rose-acacia, 74A
Rusby, 74A
Longleaf Pine, *Pinus palustris*, 26
Louisiana Palm, *Sabal louisiana*, 56

Madrone
    Arizona, 202
    Texas, 322
Magnolia
    Large-leaf, 262
    Mountain, 262
    Southern, 262A
    Swamp, 248
    Umbrella, 266
Magnolia Vine, *Schizandra coccinea*, 438A
Mahogany
    Mountain, 386
Mahonia, *Berberis spp.*, 82
Mangrove
    Black, 164
Manzanita, *Arctostaphylos pungens*, 246A
Maple
    Black, 120
    Big Tooth, 122
    Florida Sugar, 120
    Red, 124
    Rocky Mountain, 124
    Silver, 124
    Sugar, 124
Maple-leaf Viburnum, *Viburnum acerifolium*, 122
Mayten, *Maytenus texana*, 246
Mescal-bean, *Sophora secundiflora*, 92
Mesquite
    False, 100
    Honey, 66
Mexican Alder, *Alnus oblongifolia*, 300, 374
Mexican Bluewood, *Condalia spathulata* var. *mexicana*, 184
Mexican Buckeye, *Ungnadia speciosa*, 84A
Mexican Fiddlewood, *Citharexylum brachyanthum*, 156
Mexican Orange, *Choisya dumosa*, 48
Mexican Plum, *Prunus mexicana*, 382
Mexican Silktassel, *Garrya ovata*, 168
Mimosa, *Mimosa spp.*, 74A
Mistletoe, *Phoradendron spp.*, 152
Mockorange
    Hairy, 138
    Small-leaf, 162
    Thyme-leaved, 174
Montane Silverbell, *Halesia monticola*, 400
Moonseed
    Carolina, 442
    Diverse-leaf, 438
Mortonia, *Mortonia scabrella*, 192
Mountain Ash, *Sorbus dumosa* or *scopulina*, 84
Mountain Laurel, *Kalmia latifolia*, 244
Mountain Magnolia, *Magnolia pyramidata*, 262
Mountain Mahogany, *Cercocarpus montanus*, 386
Mountain Ninebark, *Physocarpus monogynus*, 310
Mountain-spray, *Holodiscus discolor*, 402
Mountain Willow, *Salix monticola*, 338
Mulberry
    French, 134
    Paper, 390
    Red, 390
    Texas, 378
Muscadine Grape, *Vitis rotundifolia*, 450
Mustang grape, *Vitis mustangensis*, 448

Myrtle Oak, *Quercus hemisphaerica,* 208, 218

Narrow-leaved Cottonwood, *Populus angustifolia,* 320
Net-leaf Oak, *Quercus rugosa,* 346
New Jersey Tea, *Ceanothus herbaceus,* 326
New Mexico Locust, *Robina neomexicana,* 74A
Ninebark
  Common, 310
  Mountain, 310
Northern Andrachne, *Andrachne phyllanthoides,* 232
Nuttall Oak, *Quercus nuttallii,* 396

Oak
  Arkansas, 242
  Arizona, 346
  Black, 312
  Blackjack, 342
  Bur, 350
  Chestnut, 344
  Chiquapin, 360
  Diamond-leaf, 236
  Durand, 212
  Emory, 354
  Gambel, 348
  Gray, 216
  Lacey, 308
  Laurel, 208
  Limestone, 218
  Live, 208A, 222
  Myrtle, 208, 218
  Net-leaf, 346
  Nuttall, 396
  Overcup, 350
  Pin, 396
  Poison, 58
  Post, 348
  Red, 312, 392
  Sandpaper, 380
  Scarlet, 396
  Shin, 368
  Shingle, 264
  Shumard's, 312, 394
  Silver-leaf, 264
  Swamp, 346
  Texas Red, 394
  Toumey, 218
  Water, 206
  Western Scrub, 368
  White, 308
  Willow, 270
Ocotillo, *Fouquieria splendens,* 6
Oiser, *Cornus spp.,* 178
Oklahoma Alder, *Alnus maritima,* 296, 388
Olive
  Desert, 128
  Wild, 178
Opposum Wood, *Halesia carolina,* 378
Overcup Oak, *Quercus lyrata,* 350
Ozark Witch-hazel, *Hamamelis vernalis,* 290, 398

Pachystima, *Pachystima myrsinites,* 130
Palm
  Dwarf, 56
  Louisiana, 56
Palo Verde, *Cercidium spp.,* 68
Panhandle Grape, *Vitis acerifolia,* 448
Paper Mulberry, *Broussonetia papyrifera,* 390
Paw-paw, *Asimina triloba,* 240
  Small, 262
Peach
  Desert, 184
Peach-leaf Willow, *Salix amygdaloides,* 296
Pecan, *Carya illinoensis,* 80
Pepper-bush
  Sweet, 404, 316
Persimmon, *Diospyros virginiana,* 274
  Texas, 266
Pickleweed, *Allenrolfea occidentalis,*

14
Pine
  Bristlecone, 24
  Limber, 24
  Loblolly, 30
  Longleaf, 26
  Pinyon, 28
  Ponderosa, 30
  Shortleaf, 28
  Slash, 30
Pinoak, *Quercus palustris*, 396
Pinyon Pine, *Pinus cembroides* or *edulis*, 28
Pipevine, *Aristolochia tomentosa*, 438A
Pistacio, *Pistacia texana*, 80
Plane-leaved Willow, *Salix planifolia*, 222
Plane-tree
  Arizona, 364
Plum
  American, 338
  Creek, 406
  Hortulan, 334
  Mexican, 383
  Wild, 360
Poinciana, *Caesalpinia gilliesii*, 96
Poison Ivy, *Rhus toxicodendron* var. *vulgaris*, 58, 418
Poison Oak, *Rhus toxicodendron* var. *quercifolia*, 58
Poison Sumac, *Rhus vernix*, 92
Polygala
  Spiny, 192
Ponderosa Pine, *Pinus ponderosa*, 30
Porlieria
  Texas, 42, 92
Possum-haw, *Viburnum nudum*, 168
Post Oak, *Quercus stellata*, 348
Post Oak Grape, *Vitis lincecomii*, 448A
Prairie Acacia, *Acacia hirta*, 104
Prairie Willow, *Salix humilis*, 220
Prickly-ash, *Zanthoxylum spp.*, 76

Quaking Aspen, *Populus tremuloides*, 296
Quinine Bush, *Cowania stansburiana* or *panicea*, 280
Rabbit-brush, *Chrysothamnus spp.*, 228, 232
Rattle Bean, *Sesbania drummondii*, or *panicea*, 104
Red Bay, *Persea borbonia*, 240
Red Beech, *Fagus grandifolia*, 384
Redbud, *Cercis canadensis*, 236
  California, 236
Red-bush, *Lippia graveolens*, 140
Red Maple, *Acer rubrum*, 124
Red Mulberry, *Morus rubra*, 390
Red Oak, *Quercus rubra* or *falcata* and hybrids, 312, 392
Rhododendron, *Rhododendron spp.*, 346
Robinia, *Robinia spp.*, 74A
Rock Elm, *Ulmus thomasi*, 376
Rocky Mountain Maple, *Acer glabrum*, 124
Rocky Mountain Willow, *Salix saximontana*, 246
Roosevelt Weed, *Baccharis neglecta*, 242
Rose, *Rosa spp.*, 72
Rose-acacia Locust, *Robinia hispida*, 74A
Rosin-bush, *Baccharis sarothroides*, 232
Rusty Viburnum, *Viburnum rufidulum*, 130

Sage
  Autumn, 166
  Blue, 140
  Greaseleaf, 134
Sage-brush, *Artemesia spp.*, 144
Sageretia, *Sageretia wrightii*, 150
St. Peter's-wort, *Ascyrum stans*, 162
  Flaxleaf, 160A
Saltbush
  Big, 254

Winged, 228
Salt-cedar, *Tamarix gallica,* 16
Sandbar Willow, *Salix interior,* 360
Sand Grape, *Vitis rupestris,* 452
Sand Holly, *Ilex ambigua,* 326
Sandpaper Oak, *Quercus pungens,* 380
Sassafras, *Sassafras albidum,* 208
Saw Greenbriar, *Smilax bona-nox,* 432
Saw Palmetto, *Serenoa repens,* 56
Sawtooth Sotol, *Dasylirion spp.,* 110
Scarlet Bouvardia, *Bouvardia ternifolia,* 156
Scarlet Oak, *Quercus coccinea,* 396
Scoular Willow, *Salix scouleriana,* 222
Senna
    Wislizenus, 90A
September Elm, *Ulmus serotina,* 288
Service-berry
    Alder-leaved, 332
    Big Bend, 386
    Utah, 328
Shadblow, *Amelanchier arborea,* 334
Shadscale, *Atriplex confertifolia,* 190
Shagbark Hickory, *Carya ovata,* 84
Shellbark Hickory, *Carya laciniosa,* 84A
Shin Oak, *Quercus havardii* and *undulata,* 368
Shingle Oak, *Quercus imbricata,* 264
Shortleaf Pine, *Pinus echinata,* 28
Shrub Bluet, *Hedyotis intricata,* 158
Shrubby Cinquefoil, *Potentilla fruiticosa,* 90
Shumard's Oak, *Quercus shumardii,* 312, 394
Siltassel
    Mexican, 168
    Wright, 168A
Silverbell
    Montane, 400
Silver Buffalo-berry, *Shepherdia argentea,* 114

Silver Maple, *Acer saccharinum,* 124
Silver-leaf Oak, *Quercus hypoleucoides,* 264
Silver-leaf Willow, *Salix argophylla,* 210
Skunk Bush, *Rhus aromatica* var. *flabelliformis,* 58
Slash Pine, *Pinus elliottii,* 30
Slippery Elm, *Ulmus rubra,* 288
Small-leaf Mockorange, *Philadelphus microphyllus,* 162
Small Paw-paw, *Asimina parviflora,* 262
Smoke Tree, *Cotinus obovatus,* 272
Smooth Alder, *Alnus serrulata,* 334
Smooth Hawthorn, *Crataegus spp.,* 186A
Smooth Indigo, *Amorpha texana* or *laevigata,* 98
Smooth Sumac, *Rhus glabra,* 86
Snake-eyes, *Phaulothamnus spinescens,* 150
Snakewood, *Colubrina texensis,* 316
Snowberry, *Symphoricarpos spp.,* 178
Snowdrop Tree, *Halesia diptera,* 384
Soapberry, *Sapindus saponaria* var. *drummondii,* 80
Soaptree Yucca, *Yucca elata,* 108
Sourwood, *Oxydendrum arboreum,* 336
Southern Arrowwood, *Viburnum dentatum,* 138
Southern Crabapple, *Pyrus angustifolia,* 186
Southern Magnolia, *Magnolia grandiflora,* 262A
Southwestern Bernardia, *Bernardia myricaefolia,* 282
Spanish Dagger, *Yucca fraxoniana, carnerosana* or *Treculeana,* 108
Spicebush, *Lindera benzoin,* 240
Spiny Polygala, *Polygala subspinosa,* 192
Spruce

Colorado Blue, 32
Englemann, 32
Stagger Bush, *Lyonia mariana*, 270
Stewartia, *Stewartia malacodendron*, 406
Sticky Baccharis, *Baccharis glutinosa*, 320
Strawberry-bush, *Euonymus americanus*, 128
Styrax, *Styrax spp.*, 222
Sugar Hackberry, *Celtis laevigata*, 236A
Sugar Maple, *Acer saccharum*, 124
Sumac
  Evergreen, 90
  Fragrant, 58A
  Little-leaf, 90A
  Poison, 92
  Smooth, 86
  Winged, 94
Summer Grape, *Vitis aestivalis*, 448A
Supplejack, *Berchemia scandens*, 438
Swamp Cottonwood, *Populus heterophylla*, 384
Swamp Magnolia, *Magnolia virginiana*, 248
Swamp Oak, *Quercus bicolor*, 346
Swamp Tupelo, *Nyssa aquatica*, 216
Sweetgum, *Liquidambar styraciflua*, 364
Sweetleaf, *Symplocos tinctoria*, 358
Sweet Mountain Grape, *Vitis monticola*, 452
Sweet Pepper-bush, *Clethra alnifolia*, 316, 404
Sweetspire, *Itea virginica*, 304
Sycamore, *Platanus occidentalis*, 364

Tamarisk, *Tamarix gallica*, 16
Tarbush, *Flourensia cernua*, 240
Texas Adolphia, *Adolphia infesta*, 6
Texas Buckthorn, *Condalia obtusifolia*, 186A
Texas Ebony, *Pithecellobium flexicaule*, 70
Texas Kidneywood, *Eysenhardtia texana*, 100
Texas Madrone, *Arbutus xalapensis*, 322
Texas Mulberry, *Morus microphylla*, 378
Texas Persimmon, *Diospyros texana*, 266
Texas Porlieria, *Porlieria angustifolia*, 42, 92
Texas Red Oak, *Quercus texana*, 394
Texas Woodbine, *Parthenocissus heptaphylla*, 418
Thick-leaved Blueberry, *Vaccinium fuscatum*, 270
Thin-brush, *Dicraurus leptocladus*, 230
Thin-leaf Hackberry, *Celtis tenuifolia*, 264
Thyme-leaved Mockorange, *Philadelphus serpyllifolius*, 174
Torrey croton, *Croton torreyanus*, 258
Toumey Oak, *Quercus toumeyi*, 218
Tree-of-heaven, *Ailanthus altissima*, 102
Trumpet Honeysuckle, *Lonicera sempervirens*, 428
Trumpet-vine, *Campsis radicans*, 412
Tulip Tree, *Liriodendron tulipifera*, 318
Tupelo
  Swamp, 216
Twinberry Honeysuckle, *Lonicera involucrata*, 424

Umbrella Magnolia, *Magnolia tripetala*, 266
Utah Honeysuckle, *Lonicera utahensis*, 426
Utah Service-berry, *Amelanchier utahensis*, 328

Viburnum
  Maple-leaf, 122
  Rusty, 130
Virginia Creeper, *Parthenocissus quinquefolia*, 418
Walnut
  Black, 86
Wahoo
  Eastern, 140
Water Elm, *Planera aquatica*, 304
Water Oak, *Quercus nigra*, 206
Wax Euphorbia, *Euphorbia antisyphilitica*, 10
Wax-myrtle, *Myrica spp.*, 316
Western Crabapple, *Pyrus ioensis*, 186
Western Erythrina, *Erythrina flabelliformis*, 74
Western Honeysuckle, *Lonicera albiflora*, 428
Western Scrub Oak, *Quercus turbinella*, 368
Western Woodbine, *Parthenocissus vitacea*, 418
White Fir, *Abies concolor*, 34
White Oak, *Quercus alba*, 308
White Willow, *Salix lasiolepis* var. *bracelinae*, 258
Whortle-berry, *Vaccinium myrtillus*, 328
Wild Olive, *Osmanthus americana*, 178
Wild Plum, *Prunus spp.*, 360
Willow
  Beaked, 210
  Black, 356
  Bluestem, 202
  Carolina, 300
  Coyote, 280A
  Desert, 144
  Dwarf, 252
  Gray-leaf, 272
  Heart-leaf, 356
  Mountain, 338
  Peach-leaf, 296

  Plane-leaved, 222
  Prairie, 220
  Rocky Mountain, 246
  Sandbar, 360
  Scoular, 222
  Silver-leaf, 210
  White, 258
  Yew-leaved, 228
Willow Oak, *Quercus phellos*, 270
Winged Elm, *Ulmus alata*, 372
Winged Saltbush, *Atriplex canescens*, 228
Winged Sumac, *Rhus copallina*, 94
Winterberry, *Ilex verticillata*, 406
Winter Fat, *Eurotia lanata*, 144
Winter Grape, *Vitis berlandieri*, 454
Wislizenus Senna, *Cassia wislizenii*, 90A
Wisteria, *Wisteria frutescens* var. *macrostachya*, 416
Witch Hazel, *Hamamelis virginiana*, 290A
  Ozark, 398
Wolfberry, *Lycium spp.*, 142
Woodbine
  Texas, 418
  Western, 418
Wooly Buckthorn, *Bumelia lanuginosa*, 114, 376
Wooly Hornbeam, *Ostrya virginiana*, 376
Wright Silktassel, *Garrya wrightii*, 168A

Yellow Honeysuckle, *Lonicera flava*, 428
Yellow-root, *Xanthorhiza simplicissima*, 42
Yellow Trumpet, *Tecoma stans*, 44
Yew-leaved Willow, *Salix taxifolia*, 228
Youpon
  Desert, 232
Yucca, *Yucca spp.*, 108
  Soaptree, 108

ABOUT THE AUTHOR

ARCHIBALD W. ROACH, Professor Emeritus of North State University, has written 35 taxonomic and ecological articles appearing in professional journals. He has received distinguished teaching awards, was a visiting lecturer at the University of Colorado where he also led an expedition in 1949 to Ungava Bay in Labrador, in charge of the boreal forest-tundra ecotone studies. Dr. Roach is also a consultant for the Gramineae, International Bureau of Taxonomy and Nomenclature in Utrecht, Netherlands. He holds a B.A. and M.A. in Botany from the University of Colorado and has received his Ph.D. in Botany from Oregon State University. He and his wife reside in Denton, Texas.